Animal Thinking

Animal Thinking

Donald R. Griffin

Harvard University Press

Cambridge, Massachusetts, and London, England   1984

This book is printed on acid-free paper, and its binding materials have been chosen for strength and durability.

Library of Congress Cataloging in Publication Data

Griffin, Donald Redfield, 1915–
  Animal thinking.

  Bibliography: p.
  Includes index.
  1. Cognition in animals. 2. Animal behavior. I. Title.
QL785.G716   1984   591.51   83-12892
ISBN 0-674-03712-X

Preface

The aim of this book is to rekindle scientific interest in the conscious mental experiences of animals. This subject was a central concern of both biologists and psychologists during the half century following the Darwinian revolution, as thoughtfully reviewed by Walker (1983), and the principal issues were thoroughly discussed by Romanes (1884) and others. But animal minds have been largely neglected by scientists since the First World War, primarily because behavioral scientists convinced themselves that there was no way to distinguish automatic and unthinking responses from behavior involving conscious choice on the animal's part. Now, however, because of impressive progress in ethology and psychology, animal thinking is again receiving serious scientific attention. A cognitive approach to ethology offers the hope that testable hypotheses can be developed, along with methods by which the thoughts and feelings of animals can be studied objectively.

A scientific advance often begins when someone surveys the unknown and outlines investigations that might reduce our uncertainty and improve our understanding. This stage is sometimes called pre-science because it anticipates the direction of scientific research, and even how it will be carried out (perhaps it requires a calculated attempt at prescience). Working scientists sometimes look down on such endeavors as useless and flimsy guesswork, but they forget that this is how it all begins. Only after we ask a question can we hope to answer it, and the importance of pre-science lies in seeking out significant questions and formulating them in ways that lead to convincing answers. Confident certainty is a luxury enjoyed by only a few areas of science, and often, as in the case of classical physics in the nineteenth century, certainty may give way to mystery as more is discovered about the real world. Conscious mental experience, in men and in animals, remains a challenging unknown territory.

This book describes and analyzes several of the more significant new developments in ethology, psychology, and neurobiology from a cognitive point of view. In advocating this approach to ethology, I am using *cognitive* in a literal sense to refer to conscious thought and knowledge, thus avoiding a recent tendency to restrict the term to information processing. This narrower usage, as in cognitive psychology, is confusing at best, and at worst it may do us a grave disservice by fostering the belief that even human thinking consists only of information processing. When the human mind is widely viewed as nothing more than a computer system, it is scarcely surprising that the possibility of consciousness in animals has been seriously neglected.

Chapters 1 and 2 present a balanced interpretation of old and new evidence about animal thinking, drawn from fields as diverse as neuroanatomy, behavioral ecology, and the philosophy of mind. The comparative perspective leads inevitably to a critical reexamination of the relationship between conscious awareness and genetically programmed behavior. This in turn calls into question the widespread assumption that only learned behavior can be accompanied by conscious thinking. Chapters 3 to 6 review numerous examples of animal behavior that can be interpreted as the outcome of conscious thinking. My emphasis is on wild animals living under natural conditions, because domesticated and captive animals are often unable to engage in normal social interactions or in the other versatile sorts of behavior that are called into play when they must cope with the challenges of life in the wild.

The final four chapters describe three areas of scientific observation and experiment that appear likely to yield better data about the existence and significance of conscious thoughts and subjective feelings. Chapter 7 outlines ways in which laboratory experiments on animal behavior, and the analysis of electrical correlates of brain activity, may be extended to detect the presence of conscious thinking and to analyze its content. Chapters 8 and 9 refine and extend my earlier suggestions that animals convey thoughts as well as feelings when they communicate with their social companions. This provides cognitive ethologists with an opportunity to literally read the minds of communicating animals. Finally, Chapter 10 discusses how consciousness may be adaptively useful to the many types of animals that live in interdependent social groups and how several cases of innovative behavior are difficult to explain as the work of thoughtless robots.

A theme that keeps recurring as one contemplates the versatility of animal behavior is the economy and efficiency of conscious thinking. Scientists have generally assumed that most animals cope with the challenges they face solely by following behavioral instructions obtained from their genetic heritage or their individual experiences. But to account for the flexibility with which many animals adapt their behavior to changing circumstances would require an enormous number of specific instructions to provide for all likely contingencies. If, on the other hand, an animal thinks about its needs and desires, and about the probable results of alternative actions, fewer and more general instructions are sufficient. Animals with relatively small brains may thus have greater need for simple conscious thinking than those endowed with a kilogram or more of gray matter. Perhaps only we and the whales can afford the luxury of storing detailed behavioral instructions, while animals with only a milligram or so of central nervous system must think consciously about their most pressing problems, if only for reasons of economy and efficiency.

It is not generally recognized how effectively the contemporary behavioristic climate of opinion in science has inhibited investigation of animals' thoughts and feelings. Throughout our educational system students are taught that it is unscientific to ask what an animal thinks or feels. Such questions are actively discouraged, ridiculed, and treated with open hostility. For example, Pribram (1978) reports that a lead article he had been invited to write for *Science* was rejected because of his inference that monkeys might be thoughtful in their approach to complex problems. Another unfortunate effect of this attitude is that field naturalists are reluctant to report or analyze observations of animal behavior that suggest conscious awareness. We have been so thoroughly brainwashed by the vehement rejection of suggestive evidence of animal thinking that it is considered foolhardy for students and aspiring scientists to let their thoughts stray into such forbidden territory, lest they be judged uncritical, or even ostracized from the scientific community.

One common objection to attempts to study animal thinking is that the only available evidence is anecdotal, and such evidence is unsatisfying to scientists because it may have been an accidental occurrence. For instance, it has occasionally been observed that an animal will give an alarm call when no danger threatens, but other animals are thereby frightened away from a morsel of food, which the first animal then seizes. Does this mean that the alarm call was given intentionally in order to secure the food? Cautious scientists tend to resist this interpretation and find single instances of such behavior unconvincing. But as Dennett (1983) has recently pointed out, "As good scientists, ethologists know how misleading and, officially, unusable anecdotes are, and yet on the other hand they are often so telling! . . . But if their very novelty and un-repeatability make them anecdotal and hence inadmissible evidence, how can one proceed to develop a cognitive case for the intelligence of one's target species?" Even the largest set of scientific data must obviously begin with datum number one, but if ethologists are deterred from pursuing the matter, additional data will never be forthcoming. The obvious remedy is to start an open-minded collection of relevant data and make every possible attempt to replicate suggestive observations.

Another sort of disparagement involves what I call the "tooth fairy reaction." If one suggests that a certain animal thinks about what it is doing, one is told that this is a childish sort of animism, comparable to naming boats or automobiles and thereby endowing them with personalities. If one asks why one's scientific colleagues are so certain that the animal is not thinking consciously, the answer often is that we cannot prove that the creature is not conscious, but then we can't really prove there is no tooth fairy either. Although it may sway the opinions of colleagues, this sort of argument is nothing more than a contorted statement of disbelief. That which one does not believe to be true is likened to something else that is generally agreed to be a fantasy. Although it is difficult to settle disagreements about the existence or content of animal consciousness, this book outlines approaches that offer some promise of success.

Philosophers often help clarify our thinking about difficult questions of this type. One can select philosophers to give support to almost any theoretical position, but in setting the stage for my inquiry into animal consciousness, the following quotations from two contemporary philosophers are appropriate: "The ultimate argument against behaviorism is simply that it seeks to prohibit *a priori* the employment of psychological explanations that may, in fact, be true" (Fodor, 1968), and "Though this behavioristic philosophy is quite fashionable at present, a theory of the non-existence of consciousness cannot be taken any more seriously, I suggest, than a theory of the nonexistence of matter" (Popper, 1974).

Many colleagues have contributed to the preparation of this book. Ruth N. Anshen stimulated its inception and its development by her emphasis on the generality and philosophical importance of the subject. Carolyn Ristau constantly tempered my enthusiasms with her understanding of the complexities of behavioral experiments and their interpretation. Patricia and Paul Churchland provided both encouragement and philosophical criticisms. Steven Hillyard and Helen Neville guided my review of electrical indices of perception in human brains and the extension of such experiments to animals. Timothy Goldsmith and Helene Jordan provided thoughtful criticisms and comments. Any errors of fact or interpretation are of course my own and have survived in spite of these constructive critics. Rosanne B. Kelly patiently produced countless drafts and revisions through many challenging interactions with a word processing system. Thoughtfully critical editing by Peg Anderson has clarified and improved almost every paragraph.

The stimulating and supportive atmosphere of The Rockefeller University made it possible for me to expand my horizons into new areas of science and scholarship, and the financial support provided by The Harry Frank Guggenheim Foundation allowed me to explore in depth the questions discussed in this book.

Contents

1 *What Do Animals Think?* *1*

2 *Other Minds* *27*

3 *Making a Living* *48*

4 *Predators and Prey* *73*

5 *Artifacts and Templates* *95*

6 *Tools and Engineering* *118*

7 *Scientific Evidence of Animal Consciousness* *133*

8 *A Window on Animal Minds* *154*

9 *Symbolic Communication* *165*

10 *Natural Psychologists* *186*

Bibliography 213

Index 229

Animal Thinking

1 *What Do Animals Think?*

What is it like to be an animal? What do monkeys, dolphins, crows, sunfishes, bees, and ants think about? Or do nonhuman animals experience any thoughts and subjective feelings at all? People have always been fascinated by the question of animal consciousness, because both pets and wild animals arouse our admiration and curiosity. They tempt us to put ourselves into their skins and imagine what their lives are like. But is this possible? Have students of animal behavior learned enough to constrain our speculations, constructively, about the thoughts and feelings of other species? If not, can they do so in the future? Most scientists concerned with animal behavior give negative answers to these questions and recoil from the prospect of mixing them into scientific thinking. But despite their strenuous objections, questions about animal awareness crop up again and again, even in the minds of scientists.

Many philosophers have recognized the deep significance of these questions, for it is a fundamental intellectual challenge to understand the nature of subjective mental experience. Conscious experiences are surely important to us, so it is of great interest to inquire whether the important class of phenomena that we call thoughts and feelings occur only in human beings. If our species is unique in this respect, what endows us with this special attribute? Is the human brain different from all other central nervous systems in some basic property that permits consciousness? Or do other animals whose central nervous systems are composed of similar neurons and synapses also think and feel? In that case how does the consciousness of other species differ from ours? This book will draw on contemporary knowledge about the realities of animal behavior in an attempt to evaluate the philosophical position expressed by David Hume (1739) that "no truth appears to me more evident, than that beasts are endow'd with thought and reason as well as men."

The contemporary philosopher Thomas Nagel has explored these questions in a perceptive article entitled "What is it like to be a bat?" (Nagel, 1974). Taking for granted that bats and other complex animals do have mental experiences, he argues that it is inherently impossible for us ever to know precisely the experiences of another species. But his argument is primarily concerned with exactness and certainty, and he seems to recognize, albeit grudgingly, that it is possible to gain some significant though partial understanding of what it is like to be a bat or any other animal. The objective of this book is to explore how an appreciation of animal behavior can help us begin to answer Nagel's question for animals in general.

Many philosophers, from Aristotle to Chomsky (1966), Popper (1972), and Davidson (1975) have insisted that true conscious experience depends upon human language. But other philosophers, such as Dennett (1978, 1983) and Bishop (1980) have advanced substantial counter arguments. The argument that there is no consciousness without language is in many ways analogous to the belief that newborn human babies are incapable of conscious thinking until they have learned by experience the properties of the world around them. But Locke's concept of the neonate human mind as a *tabula rasa* is rapidly being abandoned as a result of new discoveries about infant perception (Rose, et al., 1977; Haith, Bergman, and Moore, 1977; Jackson and Jackson, 1978; Bronson, 1982; Kuhl and Meltzoff, 1982; Mehler, 1983). The ability of newborns to imitate facial expressions (Meltzoff and Moore, 1977; Field, et al., 1982) is particularly impressive. We may be on the threshold of a comparable recognition that some nonhuman animals also enjoy mental experiences, or suffer from some of them at times.

During the twentieth century scientists have learned a great deal about the behavior of many kinds of animals under a variety of circumstances. But for the most part they have remained totally uninterested in the subjective experiences of the animals they study. Two notable exceptions are Hediger (1947, 1980) and Lorenz (1963), who have consistently endeavored to persuade their fellow ethologists that animal consciousness can be studied scientifically. In recent years some psychologists have applied to animals such terms as cognition and thought. I have urged that we reconsider the question of animal awareness, emphasizing the potential usefulness of animal communication as evidence of thinking (Griffin, 1981, 1982). In 1978 an entire issue of a scientific journal was devoted to a wide-ranging discussion of cognition and consciousness in nonhuman species (Griffin, 1978; Premack and Woodruff, 1978; Savage-Rumbaugh, Rumbaugh, and Boysen, 1978). Philosophers concerned with the nature of minds have continued to discuss the possible existence of nonhuman minds (for example, Popper, 1972; Cobb and Griffin, 1978; Bennett, 1978; Churchland, 1979; Bunge, 1980; Armstrong, 1981). Neurophysiologists and neuropsychologists have made repeated attempts to explain consciousness in terms of brain mechanisms (for instance, Eccles, 1974; Thatcher and John, 1977; Uttal, 1978; Underwood and Stevens, 1979; Davidson and Davidson, 1980; Galambos and Hillyard, 1981; Sperry, 1983). But they have been primarily concerned with human consciousness.

Most of the recent revival of interest in animal cognition has been expressed by psychologists whose background and experience has mainly involved experiments with laboratory animals. This trend is clearly evident in a symposium edited by Hulse, Fowler, and Honig (1978) and a lengthy review of that volume by Wasserman (1981). It is carried a little further in a recent symposium on animal cognition (Roitblat, Bever, and Terrace, 1983) and in a semipopular book, *Animal Thought* (Walker, 1983). In the present book I will attempt two principal extensions of and improvements on these recent reviews. First, I will consider a much wider variety of animals, with emphasis on the rich store of information about their behavior and possible thinking under natural conditions, where they cope with problems that have been crucially important to their species over the course of evolution. Second, from this naturalistic and comparative perspective, I will come to grips with the subjective aspects of animal consciousness. Under natural conditions animals make so many sensible decisions concerning their activities, and coordinate their behavior so well with that of their companions, that it has become reasonable to infer some degree of conscious thinking, anticipating, and choosing. In their social interactions some species appear to communicate their thoughts and feelings to each other. Communicative behavior offers an especially promising opportunity for ethologists to listen in and thereby gather useful information about the nature of animal consciousness.

Contemporary thinking about animal minds falls into two camps. The cognitive view is that an animal is likely to think about what it is doing, at least on some occasions. Opposed to this is the behavioristic view, which stresses the dangers of error in trying to imagine what another organism might be thinking. Strict behaviorists advocate a total disregard for subjective feelings and conscious thoughts, not only in animals but in men as well. Neither of these two viewpoints is a monolithic dogma, and both include many shades of opinion.

Despite renewed interest in animal cognition, the scientists who are willing to venture into this difficult area have tended to cling tightly to the security blanket of conventional reductionism. Although most philosophers have long since abandoned logical positivism, and many psychologists now reject the negative dogmatism of the strict behaviorists, students of animal behavior are still severely constrained by a guilty feeling that it is unscientific to study subjective feelings and conscious thoughts. Even in ethology, the study of animal behavior with emphasis on natural conditions and evolutionary influences, the ghost of Jacques Loeb (1918) still makes its cold and clammy influence felt when animal behavior is described solely in terms of stimuli, responses, and adaptive advantages. As psychologists have recognized more and more complexity and versatility in animal behavior, they have tried manfully to fit all the new knowledge into the same old pigeon holes that seemed sufficient years ago to Pavlov and Watson. For instance, Walker (1983) and several of the contributors to the symposium edited by Roitblat, Bever, and Terrace (1983) are ready to investigate animal thought or cognition but recoil in apparent horror from considering animal consciousness. Some psychologists go so far as to assert dogmatically that all animal behavior is *un*conscious. It has become respectable for scientists to speak of information processing, perception, cognition, and even thinking by nonhuman animals, but in many quarters animal consciousness remains taboo. Scientists are beginning to peek timidly around the edges of the familiar behavioristic blinders, but they are still dark-adapted, figuratively speaking, and unable to face the full glare of cognitive daylight. Very few have followed the advice of Kety (1960), "Nature is an elusive quarry, and it is foolhardy to pursue her with one eye closed and one foot hobbled."

Definitions

What do we mean by conscious thinking? Although most people do not doubt the reality of their own conscious thinking, we cannot convey to another person a complete account of all that we experience. For this reason it has been customary to argue that the words *conscious* and *think* are too vague for use in scientific analysis. I disagree with this viewpoint and am convinced that we are quite correct in our intuitive belief that conscious thinking is significant and that it has important effects on our behavior. Mental processes and subjective experiences are extraordinarily difficult to define with scientific precision, primarily because we know so little about them. As Woodfield (1976) explains, "If the term being defined is vague or open-ended, then a definition will be exact only if it too is vague and open-ended." I have attempted to draw from standard sources a set of minimal definitions that are reasonably clear and usable when we inquire about animal thinking (Griffin, 1982). The Oxford English Dictionary includes among many shades of meaning the following definitions: *conscious:* "aware of what one is doing or intending to do, having a purpose and intention in one's actions"; *consciousness:* "the totality of the impressions, thoughts, and feelings which make up a person's conscious being"; *feeling:* "pleasurable or painful consciousness, emotional appreciation or sense (of one's own condition or some external fact)"; and *think:* "to form in the mind, conceive (a thought, etc.), to have in the mind as a notion, idea, etc." These and most other dictionary definitions have a circular quality, since words are used to define each other; nevertheless we understand what they mean when applied to other people.

Armstrong (1981) argues that minds can be accounted for by a strictly materialist or physicalist philosophy, and that they are best viewed as states or processes in the central nervous system that are potentially capable of producing behavior; that is, thoughts are dispositions to behave in certain ways. But we are aware of only a small fraction of the processes occurring in our nervous systems that produce or affect our behavior. In fact, this definition of a mind would seem to include all physiological processes that initiate any sort of observable behavior in any living organism. To Armstrong's definition of minds as dispositions to behave, I would add the capability of thinking consciously about objects and events. These may be close at hand or far away in time or space. An animal may remember yesterday's tasty food or anticipate the warm dry nesting chamber at the end of its burrow, while we can imagine the Battle of Gettysburg or the first manned landing on Mars.

Armstrong goes on to define consciousness as perception of one's own mental states. He illustrates this by citing the example of a person who drives an automobile effectively but without consciously thinking about the process: "An account of mental processes as states of the person apt for the production of certain sorts of behaviour very possibly may be adequate to deal with such cases as that of automatic driving. It may be adequate to deal with most of the mental processes of animals, which perhaps spend most of their lives in this state of automatism" (p. 12). Armstrong defines true consciousness as "perception or awareness of the state of our own mind . . . a self-scanning mechanism in the central nervous system" (p. 13). It is commonly assumed that animals are incapable of knowing their own minds, if they have any minds at all. But how can we be sure that no animal is ever aware of its own mental states?

Another materialist philosopher, Bunge (1980), begins a book about the nature of minds by stating that "perceiving, feeling, remembering, imagining, willing, and thinking are usually said to be mental states or processes. (We shall ignore . . . the quaint view that there are no such facts)" (p. 1). Bunge goes farther than most philosophers in recognizing that some mammals and birds experience mental states and processes, but he asserts (pp. 74–75) that "All and only the animals endowed with plastic (uncommitted or not wired-in) neural systems are capable of being in mental states (or undergoing mental processes)" (pp. 74–75).

Bunge rigidly denies any mental life to reptiles, amphibians, and fishes and to the invertebrates, ignoring extensive data demonstrating that in many cases their behavior is plastic and versatile. Abundant evidence of learning and plasticity in a wide variety of vertebrate and invertebrate animals has been reviewed by Maier and Schneirla (1935), Thorpe (1963), Bitterman (1965), and Gould (1982). A three-volume compendium edited by Corning, Dyal, and Willows (1973–1975) provides detailed data and interpretation of learning in invertebrate animals. Specific examples of quite simple learning in isolated ganglia of mollusks have been analyzed at the cellular level by Kandel (1979a,b), and fairly complex learning in bees has been thoroughly investigated by Menzel, Erber, and Masuhr (1974); Erber (1975a,b); Menzel (1979); Erber, Masuhr, and Menzel (1980); and Klosterhalfen, Fischer, and Bitterman (1978). In later chapters I will describe other examples of versatile behavior in insects, and explain why I believe that plasticity or learning may not be necessary for conscious thinking. But in other respects the following selections, rephrased from Bunge's definitions and logical criteria, are helpful guides to our thinking about animal minds:

Mental events can cause nonmental events in the same body and conversely.   (p. 84)

We take it for granted that animals of several species know how to perform certain actions, know some constructs, and have some knowledge of events. We include among the latter the empathic knowledge of other animals. Empathy, extolled by intuitionists and mistrusted by rationalists, is admittedly fallible—but it is also indispensable.   (p. 163)

An animal . . . is aware of (or notices) stimulus x (internal or external) if and only if it feels or perceives x. It is conscious of brain process x if and only if it thinks of x. The consciousness of an animal is the set of all the states of its CNS in which it is conscious of some neural process or other in itself.   (p. 175)

An animal act is voluntary (or intentional) if and only if it is a conscious purposeful act.   (p. 182)

An animal acts of its own free will if and only if its action is voluntary, and it has free choice of its goal(s).   (p. 183)

All animals capable of being in conscious states are able to perform free voluntary acts.   (p. 184)

An animal (1) has (or is in a state of) self-awareness if and only if it is aware of itself (that is, of events occurring in itself) as different from all other entities; (2) has (or is in a state of) self-consciousness if and only if it is conscious of some if its own past conscious states; and (3) has a self at a given time if and only if it is self-aware or self-conscious at that time.   (p. 186)

To what extent can these definitions and concepts help us understand other species? The subjective experiences of nonhuman animals may be quite different from any of our own conscious thoughts or feelings, but they are probably very important to the animals concerned. A full appreciation of animals requires that we understand the nature of their thoughts.

A Materialist Approach to Mental Experience

I will take it for granted that behavior and consciousness in both animals and men result entirely from events that occur in their central nervous systems. In other words, I will operate on the basis of reductive or emergent materialism. Without attempting to resolve the philosophical mind-body problem, let me identify my approach as one that overlaps extensively with those of Churchland (1979), Bunge (1980), and Armstrong (1981). I will thus assume that there are no immaterial or supernatural processes involved in the small fraction of human or animal brain events that result in conscious subjective thoughts and feelings. This of course is only one of many philosophical positions that are intellectually tenable. One can appeal to spiritual influences as ultimately determining the nature of the physical universe but my discussion will stay out of such territory.

Because our own consciousness and thinking occur in an enormous variety of forms, these words suggest somewhat different meanings to various people. Animal thinking and feeling may also be much more varied and subtle than anything I will discuss in the following pages, but in trying to ascertain whether animals experience any conscious thoughts, it is helpful to concentrate on certain basic sorts of conscious thinking that would be the least difficult to detect in animals.

The most essential aspect of consciousness for this discussion is the ability to think about objects and events, whether or not they are part of the immediate situation. It seems likely that animals understand, at least to a limited degree under some circumstances, how their mental experiences relate to objects and events in the world around them. The animal's understanding may be accurate or misleading, and the relationships may be simple or complex. The content of conscious thinking may consist of immediate sensations, events remembered from the past, or anticipations of the future. Clearly a conscious organism must do more than merely react; it must think about something, and usually it will have some feeling about that something. Furthermore, any thinking animal is likely to guide its behavior at least partly on the basis of the content of its thoughts, so the information constituting that content must be available to the animal. Complex animals obtain most of this information through their sense organs, including those that signal conditions within their bodies, but some important information may be based on past sensory input, and some may arise through recombination into new patterns of information already present in the central nervous system.

To any biologist it will be almost self-evident that if animals do experience conscious thoughts or subjective feelings these will not be unitary all-or-nothing phenomena. Certainly our own thoughts and feelings vary enormously in their nature and complexity and in the relative importance of conscious and unconscious processes. There is every reason to expect even greater variability in other species. But before we can analyze how the content and quality of consciousness varies from species to species and from one situation to another, we must determine where and when it occurs. Animal thoughts and subjective feelings are almost certainly simpler than ours, and the content must be relevant to the animal's own situation rather than to human concerns. This makes the quest for evidence of consciousness more difficult than if we were searching for a single, well-defined entity comparable, say, to color vision. But a lack of simplicity does not render something unimportant or impossible to detect, analyze, and understand.

Neglect of Consciousness by Behavioral Scientists

Aside from Lorenz (1963) and Hediger (1947, 1968, 1980), very few ethologists have discussed animal thoughts and feelings. While seldom denying their existence dogmatically, they emphasize that it is extremely difficult, perhaps impossible, to learn anything at all about the subjective experiences of another species. But the difficulties do not justify a refusal to face up to the issue. As T. H. Savory (1959) put the matter, "Of course to interpret the thoughts, or their equivalent, which determine an animal's behaviour is difficult, but this is no reason for not making the attempt to do so. If it were not difficult, there would be very little interest in the study of animal behaviour, and very few books about it" (p. 78).

At issue here is whether animals are mechanisms and nothing more. Most biologists and psychologists tend, explicitly or implicitly, to treat most of the world's animals as mechanisms, complex mechanisms to be sure, but unthinking robots nonetheless.

Mechanical devices are usually considered to be incapable of conscious thought or subjective feeling, although it is currently popular to ascribe mental experiences to computer systems. John (in Thatcher and John, 1977), among others, has equated consciousness with a sort of internal feedback whereby information about one part of a pattern of information flow acts on another part. This may be a necessary condition for conscious thinking, but it is also an aspect of many physiological processes that operate without any conscious awareness on our part.

Many comparative psychologists seem almost literally petrified by the notion of animal consciousness. Historically, the science of psychology has been reacting for fifty years or more against earlier attempts to understand the workings of the human mind by introspective self-examination—trying to learn how we think by thinking about our thoughts. This effort led to confusing and contradictory results, so in frustration experimental psychologists largely abandoned the effort to understand human consciousness, replacing introspection with objective experiments. While experiments have been very helpful in analyzing learning and other human abilities, the rejection of any concern with consciousness and subjective feelings has gone so far that many psychologists virtually deny their existence or at least their accessibility to scientific analysis. In one rather extreme form of this denial, Harnad (1982) has argued that only after the functioning of our brains has determined what we will do does an illusion of conscious awareness arise, along with the mistaken belief that we have made a choice or had control over our behavior. The psychologists who thus belittle and ignore human consciousness can scarcely be expected to tell us much about subjective thoughts and feelings of animals. If we cannot gather any verifiable data about our own thoughts and feelings, the argument has run, how can we hope to learn anything about those of other species?

A long overdue corrective reaction to this extreme antimentalism is well under way. To a wide range of scholars, and indeed to virtually the whole world outside of narrow scientific circles, it has always been self-evident that human thoughts and feelings are real and important (see, for example, MacKenzie, 1977; and Whiteley, 1973). This is not to underestimate the difficulties that arise when one attempts to gather objective evidence about other people's feelings and thoughts, even those one knows best. But it really is absurd to deny the existence and importance of mental experiences just because they are difficult to study. Why do so many psychologists appear to ignore a central area of their subject matter when most other branches of science refrain from such self-inflicted paralysis? The usual contemporary answer to such a question is that a relatively new sort of cognitive psychology has developed during the past twenty or thirty years, based in large part on the analysis of human and animal behavior in terms of information processing (reviewed in a volume edited by Norman, 1981). Analogies to computer programs play a large part in this approach, and many cognitive psychologists draw their inspiration from the success of computer systems, feeling that certain types of programs can serve as instructive models of human thinking. Words that used to be reserved for conscious human beings are now commonly used to describe the impressive accomplishments of computers. Despite the optimism of computer enthusiasts, however, it is highly unlikely that any computer system can spontaneously generate subjective mental experience (Boden, 1979; Dreyfus, 1979; Baker, 1981).

Conspicuously absent from most of contemporary cognitive psychology is any serious attention to conscious thoughts or subjective feelings. For example, Wasserman (1983) defends cognitive psychology to his fellow behaviorists by arguing that it is not subjective and mentalistic. Analyzing people as though they were computers may be useful as an initial, limited approach, just as physiologists began their analysis of functioning of hearts by drawing analogies to mechanical pumps. But it is important to recognize the limitations inherent in this approach; it suffers from the danger of leading us into what T. H. Savory (1959) called by the apt but unfortunately tongue-twisting name of "the synechdochaic fallacy." This means the confusion of a part of something with the whole, or as Savory put it, "the error of nothing but." Information processing is doubtless a necessary condition for mental experience, but is it sufficient? Human minds do more than process information; they think and feel. We experience beliefs, desires, fears, expectations and many other subjective mental states. Many cognitive psychologists imply that a computer system that could process information exactly as the human brain does would duplicate all essential elements of thinking and feeling; others simply feel that subjective experience is beyond the reach of scientific investigation. Perhaps the issue will someday be put to an empirical test, but the extent and the complexity of information processing in our brains is so great that available procedures can detect only a tiny fraction of it, and even if it could be monitored in full detail we do not know whether any computer system could duplicate it.

The difference between conscious and nonconscious states is a significant one, yet most scientists concerned with animal behavior have felt that looking for consciousness in animals would be a futile anachronism. This defeatist attitude is based in part on convincing evidence that we do a great deal of problem solving, decision making, and other kinds of information processing without any consciousness of what is going on. Harnad (1982) bases his belief that human consciousness is merely an illusion on the fact that we are conscious of only the tip of the iceberg of information processing in our brains. Indeed the ratio of conscious to unconscious brain activity is probably even smaller than the density ratio of ice to water. The intellectual excitement of this discovery has obscured the obvious fact that we are conscious some of the time, and we certainly do experience many sorts of thoughts and feelings that are very important to us and our companions. If the choice were open, would anyone prefer a lifelong state of sleepwalking?

Cognitive Ethology

Our challenge is to venture across the species boundary and try to gather satisfactory information about what other species may think or feel. This still embryonic science of cognitive ethology should not be constrained by the computer-envy that characterizes most of contemporary cognitive psychology. We must take into account subjective experience, along with information processing, problem solving, and the survival value or evolutionary adaptiveness of animal behavior. The comparative approach to the study of mental experience may eventually prove as fruitful as it has been in anatomy, physiology, and biochemistry. If we can learn what nonhuman animals think and feel, we could base our relationships with them on factual knowledge in addition to our own emotional feelings, and at the same time we could begin to define just what is unique to our own mental life.

The consideration of animal consciousness entails a whole nest of knotty and interrelated questions. Even the preliminary steps to answering these questions require thoughtful weighing of all available and potentially obtainable evidence, rather than an appeal to prior beliefs and value judgments, which usually leads to unproductive clashes between inflexible convictions. We must be prepared to give equal weight to evidence for and against our favorite opinions, so we will find ourselves walking a sort of intellectual tightrope, balancing between two hazards. The first is to read too much into the limited evidence available, and the second is to lean so far over backward to be critically objective that we ignore or underestimate important evidence. Ideally one should walk this tightrope as a thoughtful observer or experimenter who is willing to weigh all kinds of evidence. Candid and critical review of widely diverging trends and indications will be more helpful than ringing proclamations of fervently advocated doctrine. I will not conceal my own interpretations and opinions, but I will endeavor to identify them clearly.

In the search for examples of versatile behavior it seems best to consider a broad range of species, emphasizing wild animals under natural conditions. Domestic animals and pets have been heavily influenced by their human masters, both by training and conditioning and by genetic selection for the ability to live docile lives. These effects tend to modify, sometimes drastically, the natural behavior patterns by which wild animals have become adapted for particular ways of life, either through evolutionary selection or individual learning. Laboratory animals also have been selected and conditioned to adapt to an environment that differs radically from that of their wild ancestors. The widely used laboratory animals display much less social behavior than their wild counterparts. Perhaps such behavior would complicate their care and maintenance too much, or would require too large and costly cages. Naturalistic biologists tend to feel that laboratory animals have been behaviorally emasculated. This opinion has been vigorously expressed by Hediger (1968):

Thanks to the results of breeding by man, they have undergone definite changes in the course of domestication . . . [and] . . . are artificial material for a starting point, not a natural one . . . In the case of the white rat—a domestic animal, and the favourite animal for maze experiments—a domesticated animal may be fundamentally different from its ancestral wild stock. Since the wild rat was much more difficult to handle, it was replaced in 1895 by the far more docile standardized white rat in laboratory experiments in animal psychology. Thus instead of making the experimental method fit the animal, the sort of abstract animal form, nicely adapted to the experimental apparatus, was made. (p. 10)

Dogs, cats, and riding horses often seem to know what they are doing, and much of what we know about animal behavior has been learned from studies of captive or domesticated animals. They are readily available for observation, and many variables can be better controlled by working with known individuals living under controlled conditions. But many of the situations in which thinking is most likely to emerge have been eliminated from the lives of domesticated animals. Predation is almost always lacking, although fear of it may continue. Very little effort is needed to obtain food or water; shelter is no problem and mates are ordinarily introduced at our convenience, rather than being sought out, courted, and individually selected. Human caretakers have eliminated most of the need to search for what an animal needs and of avoiding what threatens to hurt or kill it. But the way in which an animal meets these challenges may provide the best evidence about its feelings and its thoughts.

A common experimental procedure in studying animal learning is to withhold food or water until the animal behaves in a certain way and then require various further changes in its behavior. This does resemble the requirements of searching for food under natural conditions, and through such experiments we have learned much about the abilities of animals to solve problems. But experimenters ordinarily restrict the situation and the nature of the problem the animal must solve because they are looking for answers to scientific questions about learning or memory, not inquiring what the animal may be thinking or feeling.

Ethologists, behavioral ecologists, and field zoologists are devoted and diligent observers of what animals do in their normal lives. They must overcome staggering practical difficulties and endure long hours of discomfort and frustration to gather detailed and objective data. The considered judgments of experienced field naturalists are a valuable source of evidence, which scientists have not utilized adequately or effectively. We are still overreacting to the disparagement of anecdotal data, on which Darwin and Romanes relied, for lack of anything better. Indeed, field observers have been excessively inhibited by scientific critics who tend to reject as unsound all descriptive observations of animal behavior. Of course, isolated observations must be replicated and confirmed. Repeated observations, preferably documented by photographs or other objective records, are highly desirable, and it is important that new data be critically evaluated by colleagues not committed to any particular interpretation. But scientists commonly err by being too skeptical and refusing to accept evidence at its face value when it challenges established beliefs. Field observers often fail to report evidence suggestive of conscious thinking even when they obtain it, and editors of scientific journals are reluctant to publish it. These inhibitions have filtered out of the published data base many observations that could become starting points for new research.

Because this effective simplicity filter has tended to impoverish the scientific literature, and because most students of animal behavior have concentrated on quite different questions from those I am trying to answer, it is difficult to draw from their published findings a complete and balanced survey of evidence suggesting conscious thinking. We are less likely to find evidence of conscious thinking in routine repetition of stereotyped patterns than in behavior that is adjusted to changing needs and circumstances. It seems most promising to concentrate on situations in which animals successfully cope with challenging problems by choosing among alternative behavior patterns. Some species and some types of behavior have been studied thoroughly enough that we can make some inferences about thoughts or feelings that may precede or accompany it. On this basis I have selected for discussion in the following chapters foraging, predation and escape from predators, the construction and use of artifacts, and communication. Many other categories of behavior could well be considered from the same basic viewpoint, for instance, courtship, mating, rearing of young, territorial defense, dominance and aggression, habitat selection, or migration.

Sentient Nature

The environmental movement has gone a long way toward reversing what had appeared, seventy-five to a hundred years ago, to be an irresistible trend toward the elimination of natural environments and their replacement by a completely artificial world of human artifacts, lands devoted to agriculture, waterways serving for navigation and waste disposal, and, in general, the substitution of contrived for natural surroundings. This movement has assigned real and intrinsic values to animals, plants, and relatively undisturbed environments. Many of us are anxious to maintain some representative samples of rain forests, wetlands, mountains, streams, and beaches in a state approximating that which prevailed before advanced industrial society changed so much of our planet. A rarely articulated element in this concern is a feeling that many of the animals we wish to preserve may be sentient creatures whose feelings are worthy of consideration. The larger and more complex animals are an important part of what the environmental movement endeavors to protect and restore, yet there has been rather little attempt to examine why people prefer that there be an abundant, observable natural variety of wild animals. Why do we care whether we can watch deer in addition to sheep, ruffed grouse as well as chickens and turkeys, or butterflies along with cockroaches and honeybees?

Among the numerous reasons for wishing to preserve nature are our perception of kinship with other animals and our deep curiosity about their feelings and thoughts. These attitudes are reflected also in the popularity of zoos, of television programs about wild animals, and of pictures of them. Are these attitudes holdovers from an earlier stage of human history when hunting, protection against predators, and other direct interactions with a natural world were of crucial importance for human activities? Some argue that admiration of nature is merely a trivial form of recreation, of no basic importance in itself, and harmful because it impedes the fullest use of resources available for human ex-ploitation. But interest in animals is far too extensive to be dis-missed so lightly.

Another aspect of this is the widespread love of pets. Purely utilitarian reasons can scarcely justify our millions of dogs, cats, aquarium fish, and caged birds.

A major element in all this human interest in animals is a deep-seated sympathy for animals as sentient creatures. We try to un-derstand our pets or whatever wild animals we have an oppor-tunity to observe. We are challenged by the very difficulties of putting oneself in the skin of another animal, but we also are searching for empathy, a consideration that has not received nearly the attention it deserves. We like and admire other animals to a considerable extent because we enjoy trying to imagine what their lives are like *to them*. We are inclined to wonder what our pet dog wants, what the birds in our garden are feeling, or how life seems to the wide-eyed deer we glimpse from a passing auto-mobile. We feel that their ways of life must differ from ours, and similarities and differences are exciting to contemplate. Even hunting and fishing, which destroy individual animals, are often carried out by sportsmen who take an admiring interest in their quarry. Wise hunters and fishermen must also be conservationists, because they wish to continue hunting and fishing in the future.

Most of the species I have mentioned so far are mammals, birds, or other quite complex animals. But there is an enormous variety of living species, and the more one knows about them, the wider the scope for empathic curiosity becomes. We can greatly increase the potential satisfaction of learning about animals by going beyond the most familiar and locally abundant species. Practical considerations restrict most of us to studying domestic animals and the more common wild species, but the popularity of books and television programs about animals in their natural habitats attests to an insatiable appetite for broader zoological perspectives. This curiosity forms a continuum with the desire to understand people of different cultures whose lives differ from ours. In both cases it is a fundamental, and I believe significant, impulse to try to put ourselves into other shoes, or skin.

Necessary Distinctions

We would like our knowledge of animals to be accurate and realistic. Narwhals, manatees, and butterflies are more significant than unicorns, mermaids, or humbugs. Imaginary creatures have their place, but it is in the realm of fantasy where they often illuminate human perceptions. It is the scientific student of wild animals who is in the best position to inform our sympathetic curiosity, to distinguish the real from the fictional, to direct our attention toward humpback whales rather than Loch Ness monsters.

A first step for zoologists has been to survey the animal kingdom and provide accurate descriptions of all of the animals with which we share this planet, including where and how they live. This listing has been very nearly completed for the larger and most conspicuous species, although among the smaller ones, such as the insects, hundreds of new species are being discovered every year. This process of cataloguing is tedious, technical, and often frustrating. But it is important to recognize that the satisfyingly complete field guides, the accurate and often beautiful illustrations, the libraries we have come to expect for such well-known groups as birds, can only be produced by the laborious efforts of thoughtful scientists. Unfortunately we know almost nothing about the behavior of the vast majority of animal species that are already "in the books," let alone thousands more that remain to be described. Since the great majority of animal species are known only to specialized zoologists, they have only technical scientific names.

Impatience with the necessity of knowing what species one is dealing with leads many scientists specializing in other areas to deny the importance of all those tortuous taxonomic distinctions. An influential physicist, when disinhibited at a cocktail party, once berated me for what he saw as witless trivia on which biologists spend their time, "You damned biologists won't ever get anywhere until you stop this nonsense and just call 'em all bugs!" Electrons may all be identical and totally interchangeable, but even the most primitive animals are such highly organized systems that this extreme sort of simplicity filter can only lead to error and confusion. This is especially true when one studies behavior, for even closely related species or populations within species may lead very different lives and cope with differing problems, making fools of those who try to homogenize them. Furthermore, behavior often differs significantly among individuals within a population, according to sex, age, social role, health, previous experience, and perhaps even minor genetic differences. Thus if one wishes to understand the behavior of animals, or still more if one is interested in their thoughts or feelings, one must take account of their individuality, annoying as this effort may be to those who prefer the tidiness of physics, chemistry, or mathematical formulations.

Inclusive Behaviorism

Many scientists study animal behavior, but in recent decades most of their thinking, observation, and experimentation has been strongly constrained by the negative dogmatism generally known as behaviorism. This viewpoint was originally developed in American psychology under the leadership of John Watson and has been clearly summarized by B. F. Skinner (1974). The behaviorists of the 1920s were attempting to develop a rigorous science of psychology, comparable in precision to the physical and biological sciences. To do this, Watson and other behaviorists such as Skinner have insisted that only objectively observable behavior can be considered as scientific data. Because feelings, thoughts, and mental experiences cannot be observed directly, behaviorists argue that they cannot be studied scientifically. While Skinner and some other behaviorists recognize that mental experiences exist, they strongly advocate analyzing them solely in terms of their externally observable causes and results.

Our primary source of information about human thoughts, feelings, and other experiences is what people tell us by nonverbal signals, gestures, and expressions, as well as by expletives, words, and sentences. Yet twentieth-century psychology and psychiatry have shown beyond any doubt that we are aware of only a small fraction of what goes on in our brains, and we can tell other people about only a portion of what influences our behavior. Hence the behaviorists argue that introspection, the attempt to examine one's own thoughts and feelings, is a hopelessly flawed and unreliable procedure, entirely useless for scientific analysis. They follow the lead of Skinner (1957) in referring to speech as "verbal behavior." The behavioristic viewpoint has been accepted, implicitly if not explicitly, by most ethologists studying animal behavior, so it is not surprising that they learn very little about animal thoughts or feelings. Behaviorism undoubtedly had a healthy effect for many years in forcing students of animal behavior to concentrate on gathering objective evidence that could be verified by other scientists. But in the process, the subjective experiences of people and animals have come to be regarded as inconsequential fantasies.

Although behaviorism has dominated American psychology for fifty years, it has not been a monolithic orthodoxy. Edward C. Tolman (1932) called himself a purposive behaviorist and insisted that "behavior reeks of purpose and cognition." The concept that goals can govern the functioning of brains, as well as of computer systems, has come to be widely accepted, as thoughtfully reviewed by Boden (1972). Toward the end of his distinguished career Tolman (1959) acknowledged that the ethologist Otto Koehler was right to call him a "cryptophenomenologist," because he did actually believe in the reality and significance of subjective experiences or phenomena in the restricted technical meaning of the word, even though he seldom if ever stated that belief explicitly in print. Instead he concentrated on experimental evidence that animals expect certain results, especially from learned behavior. His views were not nearly so influential as those of Skinner, however, and other psychologists who insisted on ignoring mental experience. Nevertheless, many of the behavioristic inhibitions have been relaxed somewhat in recent years. A number of psychologists are beginning to gather objective, verifiable evidence about what animals may think and feel (reviewed for example by Mackintosh, 1974; and Walker, 1983). But this is still such a controversial area that animal behavior is seldom analyzed as a source of information about animal thinking.

Contemporary behaviorists often claim that they are no longer limited by the myopia of the 1920s, and in support of this claim they cite many recent studies of complex learning and problem solving in animals. But despite extensive and ingenious analysis of complicated behavior, they almost never mention the possibility that the animals might have feelings, memories, intentions, desires, beliefs, or other mental experiences. Many behaviorists continue to vehemently reject "mentalism" as though it were a deadly plague (for example, Rachlin, 1978; Wasserman, 1983). Nevertheless, many sorts of animal behavior do suggest the presence of thoughts and feelings. Hardly anyone who has lived with a pet dog or cat can seriously doubt that these animals sometimes seem to want a certain type of food, wish to be let in or out of a house, desire or dislike the companionship of particular persons or animals. Yet most behavioral scientists insist on considering only the types of learning or evolutionary selection that appear to have caused the animal to behave in a particular way. Very few scientists even realize the extent to which their thinking is constrained by this behavioristic taboo.

I suspect that animal awareness has been neglected not only for the commonly stated reason that it is difficult to study, but even more because this possibility threatens to open up a sort of Pandora's box of ideas that might contaminate science with messy and ill-defined topics. Scientists strongly prefer deterministic explanations, and the mention of animal consciousness often suggests a belief in animal free will. But the issue of determinism versus freedom of choice is only indirectly related to the question of whether nonhuman animals are conscious or not. Their conscious experiences could be the completely determined result of prior experience or genetic makeup, for example; or their choices might be free in the sense of not being predetermined, yet involve no conscious thinking.

Many scientists are convinced that even if conscious thinking does occur in animals, its existence would add nothing to a purely behavioristic analysis involving either or both of two general explanations. The first is genetic instruction, conveyed from one generation to the next by DNA in the germ plasm and molded by natural selection in the course of evolution; this sort of explanation appeals to ethologists and behavioral ecologists. The second is learning or other modification of behavior produced during an individual animal's lifetime by what Skinner has called "contingencies of reinforcement." By this he originally meant learning or, more broadly, any process whereby the favorable results of some pattern of behavior produce changes in the animal that cause this behavior to be repeated or to increase in frequency.

Skinner later combined these two influences into a larger concept of contingencies of reinforcement, which includes not only learning during an individual's lifetime, but also the effects of natural selection in "reinforcing" behavior by increasing the reproductive success of those animals in which it occurs (Skinner, 1966, 1981). While Skinner's broadened terminology is logical, it has not been widely adopted. This is unfortunate in many ways, because what we might call the neo-Skinnerian notion of inclusive contingencies of reinforcement captures an important and central aspect of the ways in which psychologists and ethologists have looked upon animal behavior. Because both groups of scientists disregard mental experiences, I will use the term "inclusive behaviorists" to designate both psychologists who are interested only in contingencies of reinforcement during an individual's lifetime, and ethologists or behavioral ecologists who are solely concerned with the effects of natural selection on behavior.

Many inclusive behaviorists argue vehemently that these two explanatory factors, acting separately or in combination, are totally sufficient for an understanding of animal behavior. This argument is often extended to include two further claims. The first is that one can predict or control the behavior of animals—and some would include people—equally well or better by considering only contingencies of reinforcement, in this neo-Skinnerian sense, without the slightest regard for subjective thoughts or feelings. The second and more basic claim is that subjective thoughts and feelings never cause or even influence behavior. If anyone doubts that behaviorists feel strongly about this point, consider the following representative statement from a recent textbook (Schwartz and Lacey, 1982): "If you want to know why someone did something, do not ask. Analyze the person's immediate environment until you find the reward. If you want to change someone's actions, do not reason or persuade. Find the reward and eliminate it. The idea that people are autonomous and possess within them the power and the reasons for making decisions has no place in behavior theory."

In another commentary Branch (1982) criticizes Roitblat (1982) for attempting to analyze the nature of representations in the brains of animals. Branch objects to what he calls "an animistic view of the behaver as originator of his actions" because, like many other behaviorists, he believes such views distract or deflect scientists from "investigation of manipulatable variables of which behavior is a function." This argument is a purely tactical one in scientific debates, as explained by Dennett (1983). It resembles in many ways the adamant insistence of many psychologists and ethologists a few years ago that it was a mistake even to consider possible genetic influences on behavior because such a concern would impede progress in analyzing the development of behavior during individual lifetimes. This viewpoint has virtually disappeared under the impact of sociobiology, and perhaps the taboos against studying mental experience will eventually fade away as well. If any scientific view has hindered investigation of important subjects, it is the behavioristic taboo against considering conscious experiences of animals and men.

This allegation that our thoughts, feelings, beliefs, plans, hopes, or desires never influence our behavior, even in the near future, is such obvious nonsense that some misunderstanding must have clouded the issue. If we take the behaviorists literally, they are telling us that we cannot anticipate or plan and that all our thoughts about what we may or may not do are irrelevant delusions. One source of confusion arises, I believe, from an implicit and often unrecognized shifting of the ground from immediately preceding causal events to what are sometimes called "ultimate" causes. The inclusive behaviorists really appear to mean something like this: thoughts or feelings and the behavior that seems to result from them must have had antecedent causes, and once these are understood one can predict the behavior without bothering about any intervening thoughts or feelings. A behavioristic psychologist will usually say that if the contingencies of reinforcement are adequately understood, the behavior can be predicted in total disregard of any subjective experiences that may occur.

This appeal to relatively remote causes often serves as an excuse for ignoring difficult problems. It is used by evolutionary biologists who shun any serious involvement with physiological mechanisms, and by ethologists and psychologists who are actively uninterested in the possibility of animal consciousness. But it is merely an excuse, because it entails a major departure from the normal interest of scientists in all important elements of a complex process. If it had been applied in other areas of biology, we would remain ignorant of many highly significant facts. For example, a biologist could deny the importance of the workings of the digestive system if he believed that all he needed to consider was the food the animal eats and the activity it fuels. If one knows that the animal's food supply is nutritious or deficient in vitamins, one can make predictions about the animal's health and vigor, even about its behavior. Why bother with the stomach, digestive enzymes, intestinal villi, or the hepatic portal system, when it is quite sufficient to recognize that food enters one side of the black box we will call an animal, and from the other side there emerges activity and metabolic byproducts? And who cares what goes on inside those messy viscera? The study of contingencies of nutrition is wholly sufficient, and it spares us the trouble of anatomical, physiological, or, more important, intellectual dissection and discovery of what really happens inside the responsive living creature.

Shallice (1972) and Sperry (1983) have argued from two quite different perspectives that conscious thinking is significant, among other reasons, because it can and does influence the function of the human brain and as a result causes overt behavior. Unfortunately neither has been able to propose just how this is accomplished in specific neurophysiological terms. Both argue for the probable correctness of our intuitive feeling that we do consider possible actions and make conscious choices between alternatives on the basis of the anticipated results. If consciousness is an illusion, as claimed by Harnad (1982), it is a remarkably useful one.

It is tempting to counter the rampant antimentalism of many psychologists by aesthetic or moral arguments about our emotional attachment to animals and our admiration of them. But I will try to avoid this temptation, without denying the importance of such aesthetic and moral considerations. Value judgments are surely enlightened by accurate understanding of the pertinent facts, and this is best achieved by putting aside our feelings as far as possible, while trying to learn what we can about the actual situation. Behaviorism should be abandoned not so much because it belittles the value of living animals, but because it leads us to a seriously incomplete and hence misleading picture of reality.

Animal Tactics

Behavioral ecologists have recently recognized that many animals employ very effective tactics in choosing a particular pattern of behavior during such activities as searching for food, courtship and mating, and doubtless in many other aspects of behavior that have not yet been studied in detail. Lloyd (1980) spoke for many of his colleagues when he recently concluded: "More often than not when you observe an insect in the field, what it is doing at the moment is the best thing it can possibly be doing at that moment for maximizing its long-run reproduction."

Behavioral ecologists use terms like choose, decide, select, search for, or avoid, but when the issue is mentioned, they go to some pains to deny that these terms imply conscious thinking. Krebs (1978) introduces a discussion of "optimal foraging: decision rules for predators" with the following cautiously agnostic disclaimer: "Note that the words 'decision' and 'choice' are not intended to imply anything about conscious thought, they are a shorthand way of saying that the animal is designed to follow certain rules" (p. 23). The rules Krebs and others have in mind are the results of evolution and natural selection. It is taken for granted that those animals which have survived and reproduced most abundantly have done so because they are superior to others of their species in structure, physiology, and behavior.

Why do hard-headed scientists, so anxious to avoid implying that animals might think or feel, use terms that in ordinary usage do connote conscious thinking? Do behavioral ecologists really mean *only* that natural selection has produced animals that behave in successful ways? Perhaps what the behavioral ecologist observes in nature suggests consciousness so strongly that part of him does wish to suggest that animals think about the likely results of their actions. Yet, catching himself using mentalistic terms, the would-be reductionist seeks qualifying disclaimers as a security blanket of scientific respectability.

The ambivalent interaction of these two tendencies is illustrated in the following quotation from a textbook of behavioral ecology by Wittenberger (1981):

Cost-benefit analyses [are discussed] *as if* behavior results from a conscious, decision-making process . . . This procedure is just a shorthand logic used for convenience. We cannot assume that animals make conscious decisions because we cannot monitor what goes on inside their heads. Nevertheless, *it really does not matter* [italics mine] what the proximate bases of those decisions are when the evolutionary reasons underlying behavior are our principal concern . . . The question of whether those choices are conscious or unconscious need not concern us, as long as we remember that tacit assumptions about purposiveness are just that . . . Particular stimuli or contexts elicit particular behaviors. An animal need not know why those stimulus-response relationships exist. It need only know what the relationships are. This knowing need not involve conscious awareness, though in many cases animals are undoubtedly conscious of what they are doing; it need only involve the appropriate neurological connections . . . Animals can be goal-directed without being purposeful, and they can behave appropriately without knowing why. (p. 48)

This assumption that through natural selection animals have arrived at nearly optimum adaptation to their environments usually entails a confident belief that behavior is influenced by genetic factors; however, this belief does not preclude the possibility that behavior is also influenced or modified by individual experience and learning. The behavioral ecologist simply assumes that natural selection has molded a genetic constitution that facilitates particular kinds of behavior, including appropriate kinds of learning. This is a plausible assumption, but it is very difficult to test, for the simple reason that evolutionary selection has operated at widely separated intervals of time over dozens or hundreds of generations in the remote past.

There is an inconsistency in assuming that evolutionary selection has produced the behavior observed in contemporary animals, yet at the same time rejecting as unscientific any suggestion that conscious thoughts and feelings might also influence what animals do. The rejection of mental experience as a causal factor is justified primarily on the ground that subjective thoughts and feelings are not observable by anyone other than the thinking or feeling creature. But the reproductive success or failure due to particular behavior patterns in remote ancestors is not observable by *anyone*. Inclusive behaviorists disparage mental explanations of animal and human behavior as unverifiable, but evolutionary selection during lengthy periods of the distant past is even less amenable to scientific scrutiny.

One response to this comparison is the claim that studies of the behavior of contemporary living animals provide reasonable evidence about how natural selection must have acted on the ancestors of these animals. Since natural conditions seldom change rapidly, today's adaptive behavior was probably also adaptive a few generations in the past. But we can compare the success of alternative patterns of contemporary behavior only in a few favorable cases, and to a limited degree, simply because almost all animal behavior really is quite efficient in solving the problems encountered in everyday life. If natural selection weeded out other, less adaptive patterns of behavior exhibited by the ancestors of today's fauna, as seems very likely, we have no data about the alternative behavior, and it is very difficult to see how this deficiency can be corrected. Nor is it possible to replicate behavioral evolution; we cannot reconstruct the past to test experimentally how the reproductive success of particular animals would have been affected had they behaved in one way or another. Thus the currently, and rightly, popular field of behavioral ecology is based on an even less testable set of hypotheses than inferences about animal thoughts and feelings derived from the study of contemporary behavior. At least the animals are behaving, and perhaps thinking, here and now. Most of their behavior can be observed repeatedly under a variety of conditions, some of which can be experimentally controlled. Indicators of their thoughts and feelings can be studied in many revealing ways.

2 *Other Minds*

Many philosophers have pondered, learnedly and intensively, what is generally called the problem of other minds. As defined by J. M. Shorter (1967), "The question of how each of us knows that there are other beings beside himself who have thoughts, feelings, and other mental attributes . . . becomes a serious and difficult [problem] because the traditional and most obvious solution to it, the argument from analogy, is open to grave objections." It is the nature of philosophers to examine important questions in excruciatingly logical detail, continually asking just what is meant by certain terms or by the thoughts that give rise to them. The common-sense view is that since other people are pretty much like me, they probably think more or less as I do, but endless arguments ensue when philosophers ask just how we can be sure of this.

An alternative viewpoint is that of the solipsist who claims there really are no minds except his own. Although the counterarguments to this absurd opinion can be weakened or demolished by a mentally agile philosopher, no one really believes he is the only sentient creature in the universe. Most objections to theories about minds other than one's own concern the difficulty of objectively verifying our inferences about other people's thoughts or feelings.

These tortuous mental exercises may seem of little or no consequence, and it is easy to become impatient while struggling to understand what philosophers are trying to tell us. But the stakes are very high, because we all make important inferences about other people's thoughts and feelings. We assume that they are happy, angry, sad, amused, affectionate, or bored; we also believe that we can judge these mental states with reasonable accuracy. Enormous difficulties arise if we misjudge what our companions are thinking and feeling, yet we must often decide what to do on the basis of inferences. Most of the time we do this quite competently. If in doubt about what our companions think or feel, we ask them, and if they are cooperative they tell us. If they feel we misread them, they ordinarily try to correct us. To be sure, such inquiries and responses are not infallible, and a whole profession of psychiatry has grown up to help us to deal with doubtful situations and to know ourselves and our fellows.

If we extend these considerations to other species, we obviously face increased difficulties. The objection that animal thoughts and feelings are inaccessible to scientific study, and that they therefore cannot be analyzed in any way, is a serious one in practical terms. But statements of this difficulty tend to be accompanied by a strong implication of total impossibility. A degree of logical rigor is demanded that is neither attainable nor expected in other areas of science. In strictly logical terms one cannot disprove solipsism, the view that I am the only conscious creature in the universe. But we ignore this quaint view because abundant, though necessarily inferential, evidence leaves no reasonable doubt that other people are also conscious creatures. Similarly we feel sure that although their thoughts and feelings may not be identical to ours, they are sufficiently similar for us to understand and appreciate them.

Many of the objections to investigation of animal thoughts and feelings seem to be based on a sort of "species solipsism." It may be logically impossible to disprove the proposition that all other animals are thoughtless robots, but we can escape from this paralytic dilemma by relying on the same criteria of reasonable plausibility that lead us to accept the reality of consciousness in other people. One argument against studying animal consciousness is that we have more than enough difficulty understanding each other, why should we trouble ourselves about the thoughts and feelings of nonhuman animals? Some inclusive behaviorists ask caustically "Who cares what animals think?" Several answers come immediately to mind. It is often important to interact with other animals, and we do so more effectively if we have some knowledge of any thoughts and feelings they are experiencing. If a wild ape's grimace expresses fear and appeasement rather than threat, this is a very important fact to understand. To be sure, direct threats of violence from other animals are rare or nonexistent for most of us. But we do interact with pets and with territorial dogs. We treat many animals as though they had thoughts and feelings, and this is a useful and effective procedure, as argued in detail by Dennett (1983). Scientists interested in animal behavior must interact with experimental animals in captivity. If a certain posture indicates that the laboratory rat is likely to bite, the scientist is well advised to learn how to read the message.

A more fundamental advantage to be gained by understanding animal thinking is that it can broaden our knowledge and appreciation of minds in general. If thoughts and feelings are limited to our species, or to some other group of animals (perhaps to birds and mammals, the so-called higher vertebrates), it would be important to learn just what enables these animals to think and feel, while all the others can only react.

Panpsychism?

One important philosophical view, sometimes called panpsychism, holds that all matter has some mental attributes, and that there is a continuum from atoms through living organisms to the human mind. An especially thoughtful version of this general perspective was developed by Whitehead (1938), who suggested that not only animals, but plants, bacteria, and nonliving matter share to some degree the essential properties that underlie thinking. This is a grossly oversimplified abstract of an immensely refined and adroitly interconnected network of penetrating ideas (reviewed in Cobb and David Griffin, 1978), but it may set the stage for inquiring whether multicellular animals need central nervous systems for conscious thinking and feeling.

Although it is generally believed that transmission, reception, and interaction of nerve impulses underlie the functioning of brains, might the information necessary for conscious thinking be provided by any other biological mechanism? One immediately thinks of the nucleic acids that store and transmit genetic instructions and pass on to cells the necessary information for the synthesis of complex protein molecules, especially enzymes that catalyze the synthesis of many important substances. Information certainly flows from DNA in chromosomes to RNA in cells and thence to enzymes that control the biochemical synthesis of myriads of essential complex molecules, but to the best of our knowledge this information flow is largely one way. To be sure, biochemical signals alter the activity of some of the nucleic acids, and usually these signals seem to come from other nucleic acid molecules. But feedback from other molecules and molecular complexes within cells also regulates cell growth and metabolic activity. Could this sort of interaction lead to conscious thinking in the complex molecular systems of even a single cell or bacterium?

This seems highly unlikely, although we cannot be certain. Since thoughts are similar to perceptions derived from sensory stimulation, it is probable that the information processing in central nervous systems is the normal, if not the only conceivable, mechanism that gives rise to conscious thoughts. All of the organisms that give evidence of experiencing such thoughts have central nervous systems consisting of the same basic elements. These are the nerve cells or neurons, the synapses that serve as junctions where excitation flows from one neuron to another, and modified neurons that synthesize and release neural transmitters, which in turn stimulate or inhibit other neurons much farther apart than the width of a synaptic gap. We should not forget that central nervous systems also contain numerous other cells, often lumped into the category of glia or supporting cells. As far as we know, they do not transmit excitation as do the neurons, but they certainly play an essential supporting, and quite probably a metabolic, role. Yet it would be unwise to close our minds to the possibility that glia may play some other major role not yet discovered.

We have no evidence that organisms lacking a central nervous system are capable of thinking about objects and events. To be sure, plants, Protozoa, bacteria, and even viruses react, grow and replicate, and all these processes require some sort of information storage and processing. Some thoughtful scholars feel that it is so preposterous to postulate conscious thinking by insects that one might as well go on to include plants, viruses, or even atoms. This seems an unjustified overreaction, but we are handicapped by our lack of any hard evidence concerning the essential processes giving rise to conscious thoughts in the human brain, where we know that they do occur. There may not be any hard and fast boundaries between the organized central nervous systems of flatworms and the nerve nets of coelenterates. Even some Protozoa, especially ciliates such as *Paramecium,* exhibit somewhat versatile behavior, as described by Jennings (1906). The organelles of ciliates function in a coordinated manner, but nothing in their makeup seems capable of storing and processing sensory information as effectively as the central nervous systems of complex multicellular animals.

Insofar as I have been able to follow Whitchcad's thinking, I believe the position outlined above is reasonably consistent with his views. The difference between coordination in ciliates and in multicellular animals is probably one of degree, without sharp qualitative discontinuities. But as Whitehead said about the difference between animals and men: "The distinction . . . is in one sense only a difference in degree. But the extent of the degree makes all the difference" (p. 38).

Another important philosophical approach is functionalism, as summarized by Churchland and Churchland (1978), and Churchland (1979, 1983). To a strict functionalist, mental states such as believing, hoping, fearing, or thinking about some object or event are defined solely by the functions they serve. The experience of fcaring that a large approaching object will cause pain consists in the functions performed by the perceived relationship, such as fleeing or hiding from the approaching object. To this extent functionalism resembles behaviorism, but functionalists attach reality and significance to the mental events themselves and to relationships among them, whereas behaviorists insist on considering only external factors impinging on the animal and any overt behavior that results. As an example of the relationship between two mental events, our hypothetical animal might combine his fear of the approaching object with a memory that yestcrday a similar object carried off one of his companions screaming with pain, tore the victim to pieces, and swallowed all but a few bits of skin. Or the fearful mental event might lead to the thought: "If I dive down my burrow, that thing won't grab me."

This sort of functionalism, which I have sketched only in a very crude form, is an obvious improvement on the negative dogmatism of the behaviorists in recognizing that some sort of internal events not only occur but have important relationships with the outside world and with each other. But functionalism as such says nothing about the nature of these internal events or relationships, as Churchland (1983) explicitly recognizes. He goes on to point out that in functional terms it matters very little just what sort of system effects a given relationship. Thus a mechanical robot controlled by a suitable computer system might exhibit the same functional relationships between impinging stimuli, its own internal states of information processing, and its motor behavior. If all these responses and relationships could be shown to be identical with those of a living animal, the strict functionalist would be obliged to conclude that there were no significant differences in the internal processes. But it does seem important to inquire whether such differences may exist and be significant. It seems plausible that the internal events that guide human or animal behavior differ greatly from those that determine the functioning of mechanical devices—even though the two types might react in the same way not only to single stimuli but to all combinations of interacting influences.

Thus functionalism does not seem adequate to account for the actual mental experiences that occur in animals. Churchland (1983) recognizes that a strict functionalist has difficulty in accounting for sensory and perceptual qualities—the redness of red, the feeling of fear or hunger—and the experience of perceiving a certain event, such as a predator killing and eating a companion. To escape from this difficulty, Churchland suggests extending strict functionalism and postulating that the neurophysiological process resulting in a particular mental experience constitutes its underlying actuality. The experience would differ according to the type of brain or other system in which a functional relationship occurs.

This elementary discussion of certain philosophical approaches to the issues of animal consciousness has barely scratched the surface of a complex intellectual fabric developed by some of the world's leading philosophers. But the functionalist approach demonstrates that what Dennett (1983) has called "the straitjacket of behaviorism" no longer constrains the imaginations of leading philosophers, if it ever did so. This should encourage us to explore animal consciousness without worrying that the effort is in any sense foolish or hopeless. It simply presents us with a series of difficult challenges, comparable, I suspect, to those confronting any scientific pioneers.

What Behavior Suggests Conscious Thinking?

In some extreme cases the likelihood is either very high or very low that thinking accompanies a given pattern of behavior. Consider for example a chimpanzee who inspects several small branches, selects one, strips it of small twigs and leaves, and carries it for a considerable distance to a termite nest. The ape then pokes this stick into termite burrows, pulls it out, and inspects it closely. Often termites are clinging to the stick, and the chimpanzee eats them with apparent relish. This behavior, graphically described by Goodall (1968, 1971), involves so many relatively disconnected steps, which differ so much from other activities of chimpanzees, that it seems obvious that the ape was thinking about collecting and eating termites while preparing the probe.

At the other extreme we can consider some of the classic cases of stereotyped behavior, often called tropisms or taxes (plural of taxis), which were studied extensively by Jacques Loeb (1918) and other biologists. Loeb selected animals and situations where behavior was relatively constant, could be described in simple terms, and could be manipulated in simple experiments. This is an entirely appropriate way to *begin* scientific analysis of a complex and puzzling set of phenomena. For example, under some circumstances many invertebrate animals will move consistently toward a light. Yet the conditions must be carefully controlled to obtain this result, and at a different temperature the same animal may move consistently away from the same light (reviewed by Maier and Schneirla, 1935). Loeb observed that if the sense organs on one side of a bilaterally symmetrical animal were more strongly stimulated, as for example by a light shining from that side, the appendages on the opposite side were more vigorously activated. This caused the animal to turn until it faced the light; the bilateral symmetry of both sensory input and motor action were restored as the caterpillar crawled toward the light. A caterpillar may move toward a light with machinelike consistency hour after hour or even day after day. This gives the investigator a satisfying feeling of understanding and control over the situation. If the animal can be caused to start and stop a given sort of behavior by simply switching on and off a light, the behavior seems to be caused in a relatively simple and straightforward way by the light. Loeb even tried to account for mammalian and human behavior along similar lines.

I have contrasted two very different kinds of animals and two very different levels of behavioral complexity. But a similar contrast can easily be made within an individual. Chimpanzees and human beings exhibit some very simple and quite predictable reflex behavior, such as the well-known knee jerk. Nevertheless, medical students must learn just how to strike a patient's knee to elicit the reflex reliably. Honeybees are strongly attracted toward light, and under many circumstances their movement toward light can be reliably and confidently predicted; at other times, however, bees communicate abstract information which they have just learned. Simplicity or complexity of behavior is not rigidly linked to particular kinds of animals. Although some species are more likely to be versatile and adaptable, there is enormous variation, and both simple and complex patterns may be exhibited by the same animal under different conditions.

Just what is it about some kinds of behavior that leads us to feel that it is accompanied by conscious thinking? Comparative psychologists and biologists worried about this question extensively around the turn of this century. No clear and generally accepted answers emerged from their thoughtful efforts, and this is one reason why the behavioristic movement came to dominate psychology. Complexity is often taken as evidence that some behavior is guided by conscious thinking. But complexity is a slippery attribute. One might think that simply running away from a frightening stimulus was a rather simple response, yet if we make a detailed description of every muscle contraction during turning and running away, the behavior becomes extremely complex. But, one might object, this complexity involves the physiology of locomotion; what is simple is the direction in which the animal moves. If we then ask what sensory and central nervous mechanisms cause the animal to move in this direction, the matter again becomes complex. Does the animal continuously listen to the danger signal and push more or less hard with its right or left legs in order to keep the signal directly behind it? Or does it head directly toward some landmark? If the latter, how does it coordinate vision and locomotion? Again one might say that the direction of motion is simple, and it is irrelevant to worry about the complexities of the physiological mechanisms involved.

But how is this simple direction "away from the danger" represented within the animal's central nervous system? Does the animal employ the concepts of *away from* and *danger*? If so, how are such concepts established? Even though we cannot answer the question in neurophysiological terms, it is clear that running away from something is a far simpler behavior than a chimpanzee's preparation and use of a probe for termite fishing. Conversely, even the locomotor motions of a caterpillar moving toward a light are not simple when examined in detail. What is simple is the abstract notion of *toward* or *away,* but the mechanistic interpretation of animal behavior tends to deny that the animal could think in terms of even such a simple abstraction.

One very important attribute of animal behavior that seems intuitively to suggest conscious thinking is its adaptability to changing circumstances. If an animal repeats some action in the same way regardless of the results, we assume that a rigid physiological mechanism is at work, especially if the behavior is ineffective or harmful to the animal. When a moth flies again and again at a bright light or burns itself in an open flame, it is difficult to imagine that the moth is thinking, although one can suppose that it is acting on some thoughtful but misguided scheme. When members of our own species do things that are self-damaging or even su-icidal, we do not conclude that their behavior is the result of a mechanical reflex. But to explain the moth flying into the flame as thoughtful but misguided seems far less plausible than the usual interpretation that such insects automatically fly toward a bright light, which leads them to their death in the special situation where the brightest light is an open flame.

Conversely, if an animal manages to obtain food by a complex series of actions that it has never performed before, intentional thinking seems more plausible than rigid automatism. In England in the 1930s some great tits and other small chickadee-like birds discovered that they could obtain cream by pecking through the aluminum foil covering milk bottles, as discussed in Chapter 3. When the birds were learning about this new food source, it seems likely that they were intentionally seeking out milk bottles on doorsteps in the early morning and pecking through their shiny coverings with the conscious intention of obtaining food.

Another criterion upon which we tend to rely in inferring con-scious thinking is the element of interactive steps in a relatively long sequence of appropriate behavior patterns. Effective and versatile behavior often entails many steps, each one modified according to the results of the previous actions. In such a complex sequence the animal must pay attention not only to the immediate stimuli, but to information obtained in the past. Psychologists once postulated that complex behavior can be understood as a chain of rigid reflexes, the outcome of one serving as stimulus for the next. Students of insect behavior have generally accepted this explanation for such complex activities as the construction of elaborate shelters or prey-catching devices, ranging from the un-derwater nets spun by certain caddis-fly larvae to the magnificent webs of spiders. But the steps an animal takes often vary, depending on the results of the previous behavior and on many influences from the near or distant past. The choice of *which* past events to attend to may be facilitated by conscious selection from a broad spectrum of memories.

In many cases these networks of informative events are sufficiently complicated that we are not sure what the animal is doing even when we know most of the relevant facts. Consider, for example, how certain ground-nesting birds lead predators away from their nests or young. This topic will be discussed in more detail in Chapter 4, but here it is enough to point out that the type of behavior depends on the kind of predator, its proximity to the young, whether the young are eggs or mobile nestlings, whether the predator has responded to the bird's previous actions, and sometimes what the other parent has been doing. Killdeer have been observed to use very different tactics when their nests are approached by cattle, which may trample on eggs or nestlings but will not eat them. Rather than moving away from the nest and fluttering as though injured, the birds stand close to the nest and spread their wings in a conspicuous display that usually causes the cattle to step aside (Skutch, 1976).

One further consideration can help refine these criteria. We can easily change back and forth between thinking consciously about our own behavior and not doing so. When we are learning some new task such as swimming, riding a bicycle, driving an automobile, flying an airplane, operating a vacuum cleaner, caring for our teeth by some new technique recommended by a dentist, or any of the large number of actions we did not formerly know how to do, we think about it in considerable detail. But once the behavior is thoroughly mastered, we give no conscious thought to the details that once required close attention. This change can also be reversed, as when we make the effort to think consciously about some commonplace and customary activity we have been carrying out for some time. For example, suppose you are asked about the pattern of your breathing, to which you normally give no thought whatsoever. But you can easily take the trouble to keep track of how often you inhale and exhale, how deeply, and what other activities accompany different patterns of breathing. You can find out that it is extremely difficult to speak while inhaling, so talking continuously requires rapid inhalation and slower exhalation. This and other examples that will readily come to mind if one asks the appropriate questions show that we can bring into conscious focus activities that usually go on quite unconsciously.

The fact that our own consciousness can be turned on and off with respect to particular activities tells us that in at least one species it is not true that certain behavior patterns are always carried out consciously while others never are. It is reasonable to guess that this is true also for other species. Well-learned behavior patterns may not require the same degree of conscious attention as those the animal is learning how to perform. This in turn means that conscious awareness is more likely when the activity is novel and challenging; striking and unexpected events are more likely to produce conscious awareness.

Thus it seems likely that a widely applicable, if not all-inclusive, criterion of conscious awareness in animals is *versatile adaptability of behavior to changing circumstances and challenges*. If the animal does much the same things regardless of the state of its environment or the behavior of other animals nearby, we are less inclined to judge that it is thinking about its circumstances or what it is doing. Consciously motivated behavior is more plausibly inferred when an animal behaves appropriately in a novel and perhaps surprising situation that requires specific actions not called for under ordinary circumstances. This is a special case of versatility, of course, but the rarity of the challenge combined with the appropriateness and effectiveness of the response are important indicators of thoughtful actions. There are limits to the amount of novelty with which a species can cope successfully, and this range of versatility is one of the most significant measures of mental adaptability. This discussion of adaptable versatility as a criterion of consciousness implies that conscious thinking occurs only during learned behavior, but we should be cautious in accepting this belief as a rigid doctrine, as will be discussed in more detail later in this chapter.

Another aspect of conscious thinking is anticipation and intentional planning of an action with conscious awareness of its likely results. Anticipation and planning are of course impossible to observe directly in another person or animal, but indications of their likelihood are often observable. As early as the 1930s Konrad Lorenz studied the intention movements of birds (Lorenz, 1971), and other ethologists had noted that these movements, small-scale preliminaries to major actions such as flying, often serve as signals to others of the same species. Although Lorenz interpreted the movements as indications that the bird was planning and preparing to fly, the term intention movement has been quietly dropped from ethology in recent years. I suspect this is because the inclusive behaviorists fear that the term has mentalistic implications. Earlier ethologists such as Daanje (1951) described a wide variety of intention movements in many kinds of animals, but their interest was in whether the movements had gradually become specialized communication signals in the course of behavioral evolution. The possibility that intention movements indicate the animal's conscious intention has been totally neglected by ethologists during their behavioristic phase, but we may hope that the revival of scientific interest in animal thinking will lead cognitive ethologists to study whether such movements are accompanied by conscious intentions. The very fact that intention movements so often evolve into communicative signals may reflect a close linkage between thinking and the intentional communication of thoughts from one conscious animal to another.

These considerations lead us directly to a recognition that because communicative behavior, especially among social animals, often seems to convey thoughts and feelings from one animal to another, it can tell us something about animal thinking. As discussed more fully in Chapters 8 and 9, it can be an important "window" on the minds of animals. Human communication is hardly limited to formal language; nonverbal communication of mood or intentions also plays a large and increasingly recognized role in human affairs. We make inferences about people's feelings and thoughts, especially those of very young children, from many kinds of communication, verbal and nonverbal; we should similarly use all available evidence in exploring the possibility of thoughts or feelings in other species. When animals live in a group and depend on each other for food, shelter, warning of dangers, or help in raising the young, they need to be able to judge correctly the moods and intentions of their companions. This extends to animals of other species as well, especially predators or prey. It is important for the animal to know whether a predator is likely to attack or whether the prey is so alert and likely to escape that a chase is not worth the effort. Communication may either inform or misinform, but in either case it can reveal something about the conscious thinking of the communicator.

Inclusive behaviorists insist on limiting themselves to stating that the animal benefits from accurate information about what the other animal will probably do. But within a mutually interdependent social group, an individual can often anticipate a companion's behavior most easily by empathic appreciation of his mental state. The inclusive behaviorists will object that all we need postulate is behavior appropriately matched to the probabilities of the companions *behaving* in this way or that—all based on contingencies of reinforcement learned from previous situations or transmitted genetically. But empathy may well be a more efficient way to gauge a companion's disposition than elaborate formulas describing the contingencies of reinforcement. All the animal may need to know is that another is aggressive, affectionate, desirous of companionship, or in some other common emotional state. Judging that he is aggressive may suffice to predict, economically and parsimoniously, a wide range of behavior patterns depending on the circumstances. Neo-Skinnerian inclusive behaviorists may be correct in saying that this empathy came about by learning, for example, the signals that mean a companion is aggressive. But our focus is on the animal's possible thoughts and feelings, and for this purpose the immediate situation is just as important as the history of its origin.

Humphrey (1976) has extended an earlier suggestion by Jolly (1966) that consciousness arose in primate evolution when societies developed to the stage where it became crucially important for each member of the group to understand the feelings, intentions, and thoughts of others. When animals live in complex social groupings, where each one is crucially dependent on cooperative interactions with the others, they need to be "natural psychologists," as Humphrey puts it. They need to have internal models of the behavior of their companions, to feel with them, and thus to think consciously about what the other one must be thinking or feeling. Following this line of thought, we might distinguish between the animals' interactions with some feature of the physical environment or with plants, on the one hand, and interactions with other reacting animals, usually their own species, but also predators and prey, as discussed in Chapter 4. While Humphrey has so far restricted his criterion of consciousness to our own ancestors within the past few million years, it could apply with equal or even greater force to other animals that live in mutually interdependent social groups.

All this adds up to the simple idea that when animals communicate to one another they may be conveying something about their thoughts or feelings. If so, eavesdropping on the communicative signals they exchange may provide us with a practicable source of data about their mental experiences. When animals devote elaborate and specifically adjusted activities to communication, each animal responding to messages from its companion, it seems rather likely that both sender and receiver are consciously aware of the content of these messages.

The Adaptive Economy of Conscious Thinking

The natural world often presents animals with complex challenges best met by behavior that can be rapidly adapted to changing circumstances. Environmental conditions vary so much that for an animal's brain to have programmed specifications for optimal behavior in all situations would require an impossibly lengthy instruction book. Whether such instructions stem from the animal's DNA or from learning and environmental influences within its own lifetime, providing for all likely contingencies would require a wasteful volume of specific directions. Concepts and generalizations, on the other hand, are compact and efficient. An instructive analogy is provided by the hundreds of pages of official rules for a familiar game such as baseball. Once the general principles of the game are understood, however, quite simple thinking suffices to tell even a small boy approximately what each player should do in most game situations.

Of course, simply thinking about various alternative actions is not enough; successful coping with the challenges of life requires that thinking be relatively rapid and that it lead both to reasonably accurate decisions and to their effective execution. Thinking may be economical without being easy or simple, but consideration of the likely results of doing this or that is far more efficient than blindly trying every alternative. If an animal thinks about what it might do, even in very simple terms, it can choose the actions that promise to have desirable consequences. If it can anticipate probable events, even if only a little way into the future, it can avoid wasted effort. More important still is being able to avoid dangerous mistakes. To paraphrase the philosopher Karl Popper (1972), a foolish impulse can die in the animal's mind rather than lead it to needless suicide.

I have suggested that conscious thinking is economical, but many contemporary scientists counter that the problems mentioned above can be solved equally well by unconscious information processing. It is quite true that skilled motor behavior often involves complex, rapid, and efficient reactions. Walking over rough ground or through thick vegetation entails numerous adjustments of the balanced contraction and relaxation of several sets of opposed muscles. Our brains and spinal cords modulate the action of our muscles according to whether the ground is high or low or whether the vegetation resists bending as we clamber over it. Little, if any, of this process involves conscious thought, and yet it is far more complex than a direct reaction to any single stimulus.

We perform innumerable complex actions rapidly, skillfully, and efficiently without conscious thought. From this evidence many have argued that an animal does not need to think consciously to weigh the costs and benefits of various activities. Yet when we acquire a new skill, we have to pay careful conscious attention to details not yet mastered. Insofar as this analogy to our own situation is valid, it seems plausible that when an animal faces new and difficult challenges, and when the stakes are high— often literally a matter of life and death—conscious evaluation may have real advantages. Inclusive behaviorists often find it more plausible to suppose that an animal's behavior is more efficient if it is automatic and uncomplicated by conscious thinking. It has been argued that the vacillation and uncertainty involved in conscious comparison of alternatives would slow an animal's reactions in a maladaptive fashion. But when the spectrum of possible challenges is broad, with a large number of environmental or social factors to be considered, conscious mental imagery, explicit anticipation of likely outcomes, and simple thoughts about them are likely to achieve better results than thoughtless reaction. Of course, this is one of the many areas where we have no certain guides on which to rely. And yet, as a working hypothesis, it is attractive to suppose that if an animal can consciously anticipate and choose the most promising of various alternatives, it is likely to succeed more often than one that cannot or does not think about what it is doing.

Conscious Instincts?

Romanes (1884) defined instinct as "reflex action into which there is imported the element of consciousness." But his views on this subject have been largely discarded in the twentieth century, principally because he has been accused of uncritically accepting too much anecdotal evidence in favor of the conclusion that animals sometimes think consciously about what they do. In recent times it has been taken for granted that only learned behavior could possibly be accompanied by conscious thinking. Most scientific discussions of conscious thoughts and feelings in animals have relied on evidence of learning. Thus it has been argued, by Bunge (1980) and many others, that only animals that can learn to adapt their behavior to changing circumstances can have conscious thoughts. For example, the "cryptophenomenologist" E. C. Tolman (1932, 1959) advocated "purposive behaviorism," by which he meant that in some sorts of behavior the animal is aware of a goal that it is seeking to attain. He believed that when a laboratory rat had just learned to solve a difficult problem, it consciously understood what it had done. To be sure, Tolman seldom used the dirty word *consciousness*, but he clearly believed that the rat experienced some sort of subjective awareness at these moments of discovery. Tolman's inference may well apply to other animals, and it may be that in many animals conscious thinking helps the animal figure out what it must do to get food or some other reward.

In contrast to learned behavior, much of which is relatively complex, most of our physiological functions proceed smoothly without our conscious awareness. These functions have reached this efficient state through the growth and integration of millions of cells following the genetic instructions transmitted from one generation to the next by DNA. Having recognized this and having learned that much of animal behavior is almost wholly under genetic control, we have concluded that it must lack any accompanying consciousness. For instance, many insects and spiders carry out quite elaborately integrated patterns of behavior, and they do so almost perfectly on the first appropriate occasion, without any opportunity to learn what to do. This absence of learning is then taken, almost universally, as proof that the animal has no conscious awareness of its instinctive behavior.

But perhaps we should pull back for a moment and ask ourselves just what evidence supports this deep-rooted assumption, which arises, I suspect, from analogies to our own situation. Human lives clearly require an enormous amount of learning, so much so that many have denied the existence of instinctive, genetically programmed human behavior. It is widely believed that only the simplest human reactions such as eye blinks, knee jerks, sneezing, cries of pain, exclamations when startled, or a newborn baby's suckling, are under predominantly genetic control. Many of these reactions happen automatically, unintentionally, and without any learning, although we may be aware of them as they occur. We do not plan to sneeze although we certainly know we are sneezing. But we may not even realize that we have blinked in response to a flash of light or the sudden sight of something moving rapidly toward us. From these experiences we reason that when animal behavior requires no learning it cannot be accompanied by conscious thought.

Consciousness of one's bodily activities falls into two general categories: we may consciously anticipate, plan, and intend to perform some action, or our bodies may simply do something without any conscious expectation and perhaps without our being able to affect the action. Yet even in cases of the second type we may be completely conscious of what our body is doing. A typical case of the first type is reaching out to grasp something; this requires consciously deciding to grasp the object, although through absent-mindedness or under hypnosis it may occur involuntarily. The second category might be exemplified by the withdrawal response to a painful stimulus. Both of these types of conscious awareness of bodily actions may occur in nonhuman animals.

Simple human reflexes have served as our "type specimens" of instinctive behavior because it is so difficult to ascertain which, if any, of our own more complex reactions have an essentially genetic basis. But the simple cases have tended to color our picture of unlearned behavior as a whole. Perhaps we are being unwisely anthropocentric in assuming that this picture accurately describes instinctive behavior in animals. When an animal carries out complex behavior, such as prey capture or nest building, without having had any opportunity for learning, we assume that it must be largely determined by genetic instructions. To be sure, it is extremely difficult to tease out the relative importance of the genetic and experiential components in a given pattern of behavior. For instance, recent ethological studies have shown that species specificity, the near constancy of a given behavior among all members of a species, does not necessarily mean that no learning is involved. In the few cases where the evidence is reasonably complete and satisfactory, it seems clear that the genetic instructions are rather general and that the individual animal learns the details.

The large genetic component in many sorts of animal behavior does not justify the conclusion that all instinctive behavior is a homogeneous category. In particular, the analogy to our own situation does not establish how tightly consciousness is linked to learned behavior rather than hereditary constitution. The suggestion that an animal, or even a person, might be conscious of his instinctive behavior is unfamiliar, and many will doubtless find it troubling. But why is this possibility so unpalatable? Before rejecting it outright, let us inquire whether this concept might help us make sense out of the confusing relationship between behavior and consciousness. We may not intend to sneeze or flinch, indeed we may be quite unable to avoid doing so, but in some cases we certainly know it is happening. When animals behave instinctively, might they be fully aware of what they are doing, without understanding the causes of their behavior or its future consequences? I will return to this line of inquiry in later chapters.

Another fundamental point requires clarification before we proceed to specific examples of animal behavior that suggest conscious thinking. If we accept the basic materialistic assumptions expressed in Chapter 1, it follows that people's conscious thoughts and subjective feelings are caused by some series of events in their brains. Although we cannot rigorously prove that all the critically causal events take place in the central nervous system, everything we know about neurophysiology points in that direction, even though a normally functioning brain must operate in close harmony with the rest of the body. Kidneys, arteries, and adrenal glands are also necessary for consciousness, but in a supporting rather than a crucial role. We know next to nothing about how brain functions that do lead to human consciousness are distinguished from those that do not. Although we can make general inferences that certain parts of the human brain, such as the cerebral cortex or reticular system, are more important than others for conscious thought, all known structures and functions of nerve cells seem to be much the same wherever we find them, whether in different parts of the human brain or in other brains. It seems very unlikely that there are "consciousness neurons" or specific biochemical substances, perhaps neurotransmitters, that are uniquely correlated with the conscious state, so that a person is conscious when and only when these cells are active or these substances are present. It seems far more likely that consciousness results from patterns of activity involving thousands or millions of neurons.

Could these patterns arise in conjunction with unlearned, genetically programmed behavior? If activation of a certain set of neurons in an appropriate pattern produces a particular thought or feeling, might this pattern of activation arise internally as a result of DNA-guided growth and cellular activity of the central nervous system itself? Asking this question makes it clear how little basis we have for any confident answer. We know too little about the neurophysiological basis of consciousness to predict whether consciousness-producing processes can appear only in conjunction with learned behavior. As with so many other topics considered in this book, it is best to recognize our ignorance and refrain from dogmatic assertions, despite the comforting familiarity of the confident assumption that unlearned behavior never entails consciousness.

Our own conscious thoughts need not be tightly linked to any overt behavior at all. We can think about objects or events, including past or future activities, without doing anything. It can be argued that our previous learning has led indirectly to the unexpressed thought, yet we can certainly have conscious thoughts that are unrelated to any current behavior or sensory input. Recognizing this obvious fact, we can ask whether conscious thoughts might sometimes arise as a result of the brain's genetically guided development and functioning. Might a brain attain the state necessary to produce conscious thought without prior stimulation by any sensory input comparable to the content of the thought?

Let us consider a specific example. Suppose the genetic instructions leading to the development of a normal brain call for a startle reaction, and the subjective emotion of fear, when a certain pattern of stimulation is received. This might be something very general, such as a large conspicuous object approaching at high speed, or it might be more specific, for example, a scowling face. Let us further assume that the pattern of reaction to this stimulus is in no sense learned, and that it appears full blown in people or animals never previously exposed to anything of the kind. If the emotion of fear and the behavior of flinching or fleeing occur consistently and predictably, it would be reasonable to infer that both the feeling and the instinctive behavior resulted primarily from genetic influences. These are difficult points to establish, because a wide variety of subtle influences from the environment can affect the development of both the nervous system and the behavior. But if the reaction is intense and highly specific, and if the pattern of stimulation is one that we can be sure the animal never encountered previously, the presumption of genetic origin becomes very strong indeed.

The traditional view has been that instinctive behavior cannot be accompanied by conscious thought. But in many cases the response and the physiological signs of emotional arousal seem very similar to what occurs during and after a learned response of the same sort. Suppose an animal instinctively retreats from, and is frightened by, large approaching objects, and further suppose that through intensive training the animal learns that a large approaching object signals the arrival of food, but that a rapidly *receding* object of the same sort signals painful punishment. After thoroughly learning about this new state of affairs, the animal would rush to greet the approaching food source and flee in terror from the same object when it was moving away. If we could convincingly demonstrate all of this, would we have to infer that the animal is conscious of its learned but not of its unlearned responses? It seems likely that conscious thinking and subjective feelings could accompany both sorts of response.

But must there be any overt response at all for a brain to attain the state necessary for conscious thought? We know that we can think about and be emotionally aroused by memories of past events without any overt motor behavior. If the same brain state could result from genetic instruction, it would produce similar thoughts and emotions. This line of thought strays even farther than the rest of this book from conventional scientific assumptions, but it is important because, if correct, this would imply the existence of genetically programmed thoughts and feelings. Are we converging on the philosopher's concept of innate ideas? Only in a restricted sense, because in most such discussions philosophers confine themselves to relatively complex human ideas and indeed seem to exclude a priori any possibility that ideas could occur in other species. But further development, and possibly empirical confirmation, of these suggestions could contribute importantly to our general understanding of thoughts and feelings.

Jung (1973) and other psychiatrists have felt that certain basic notions are so deeply imbedded in the human unconscious as to indicate an evolutionary basis, which suggests that some of these notions could be shared by other species. This intellectual water is deeper and murkier than the shallows where I am cautiously wading, but such speculation might be clarified by recognizing that the customary equating of instinctive with unconscious may be quite unsound. If we find that other species do experience conscious thoughts, we may be able to bridge the human–animal distinction in constructive ways that can enlighten even such subjects as psychiatry.

Where does all this leave us? If learning is not a reliable criterion of consciousness, what criteria can we use instead? Unfortunately, none of the criteria discussed above, individually or taken together, provide a foolproof litmus test for conscious awareness. My own inclination is to emphasize versatility and adaptedness to the problems animals face in the natural world where their species has evolved. When animals adjust their behavior effectively to solve problems, I surmise that they are likely to think and feel consciously to some degree, whether the successful behavior came about through natural selection or through learning. But this opinion can be neither confirmed nor refuted by the evidence now available to us.

Does this mean that it is pointless to consider these questions? I think not, because human consciousness and subjective feeling are so obviously important and useful to us that it seems unlikely that they are unique to a single species. This assumption of a human monopoly on conscious thinking becomes more and more difficult to defend as we learn about the ingenuity of animals in coping with problems in their normal lives. The following chapters will not prove or disprove the existence of animal consciousness, but they will review some types of behavior where simple thinking would seem to be very helpful to the animals and to their survival and successful reproduction.

3 *Making a Living*

In the natural world food is seldom available in such abundance that an animal can satisfy its needs without considerable effort. To be efficient, these efforts require something remarkably like thinking. It would be absurdly maladaptive for an animal to dash about at random, opening its mouth hopefully but without good reason to suppose there is food to be seized. Hungry animals seek food in places where they have learned it is likely to be available, and their searching behavior varies greatly, depending upon the kind of animal and the environment in which it lives. Earthworms burrowing through the soil can hardly employ tactics similar to those of a hawk soaring high in the air and scanning the ground for living prey.

This chapter will describe a number of animals whose food-gathering behavior has been studied in sufficient detail to provide a reasonably clear picture of the problems they face and the solutions they achieve. Most wild animals spend considerable time either feeding or searching for food, but much more careful and analytical data gathering is necessary before we can understand their tactics of food gathering and the thoughts they may have as they search for food.

Some very simple behavior patterns have been suggested as sufficient to account for the food gathering of certain animals.

Earthworms eat their way through the soil, swallowing the dirt they encounter and passing out of the anus a large fraction of this material. Their digestive tract breaks down the small living organisms and fragments of dead animals and plants, and molecules of carbohydrates, fats, and amino acids are absorbed into the bloodstream through the wall of the intestine. In the scientific climate that places a high value on simplicity, it is supposed that this is all an earthworm need do: open mouth, wriggle forward forcing soil into the stomach, defecate material that has not been digested and absorbed, and continue indefinitely.

But earthworms also feed in many other ways, as described a hundred years ago by Charles Darwin (1882). Even when they feed by swallowing the soil, not all soils are equally rich in digestible materials. Every fisherman who digs for earthworms to use as bait knows that some areas of soil contain many more worms than others. When an earthworm finds the soil unsuitable, either because it is too hard or because it contains very little food, it may well look for more productive areas. Its efficiency would be improved over purely random burrowing and swallowing by even so simple a tactic as, "If you can't feed, turn." This sort of explanation has been the subject of rather elaborate quantitative descriptions, from Loeb (1918) and Fraenkel and Gunn (1940) to Schöne (1980), which seem to account for the chemical responses of bacteria (Adler, 1976). But it is clearly mistaken to think that earthworms behave *only* in this simple way.

Earthworms commonly crawl out of their burrows at night to feed on the leaves of certain plant species, selected presumably on the basis of chemical senses. The earthworm grasps part of the leaf, for example the tapered tip opposite the petiole, then pulls the leaf back into the burrow to be eaten underground. Sometimes a leaf is left partly protruding from the mouth of the burrow, but at other times, either after the leaf has been pulled underground or when the worm has been on the surface for other reasons, the worm may plug the opening of the burrow with earth (Edwards and Lofty, 1972). These complexities, to which ethologists ordinarily pay little attention, demonstrate how easy it is to overemphasize the most easily explained aspects of animal behavior. Scientists so strongly prefer simplicity that their written descriptions and our resulting picture of animal behavior has passed through an effective simplicity filter and is thereby quite distorted.

Another example is provided by the baleen whales, which are said to feed in a simple fashion, almost like the ingestive burrowing of earthworms. They open their enormous mouths, take in a large volume of water, then force the water out through the narrow passages between the sieve-like plates of baleen, which trap the crustaceans or fish. This description sounds so simple that one can imagine a whale obtaining all the food it needs simply by swimming about, opening and closing its mouth. But the ocean is not supplied with a uniformly high density of small marine organisms. Rather little is known about the distribution of local concentrations of fish or krill or how whales locate them, but we do know that the baleen whales often swim for long distances between one season and the next, probably to reach areas where food is plentiful.

One of several specialized feeding tactics of humpback whales has recently received particular attention (Jurasz and Jurasz, 1979; Herman, 1980; Hain et al., 1982). The whales rise slowly toward the surface while exhaling a stream of fine air bubbles, not randomly but in definite circular patterns. The result is an effective net of bubbles, which small fish and invertebrates seem reluctant to swim through. As the fish turn back from the bubbles, they are concentrated where the whale can gather a large mouthful. The coordination of swimming and controlled exhalation necessary for this type of feeding is certainly not a simple reflex action. The whale must first discover where there are enough planktonic animals to make the effort worthwhile, then dive to a suitable depth and swim slowly upward while exhaling bubbles in a coordinated pattern. Sometimes a group of humpback whales seems to cooperate by approaching a school of fish in an effectively coordinated attack.

Earthworms and humpback whales obviously differ enormously in size and complexity of the central nervous system. We know next to nothing about how the bubble net feeding behavior develops in young humpback whales or how it varies from time to time and place to place. We have only barely sufficient evidence to conclude that fish and krill are concentrated inside the bubble net just before the whale reaches the surface. To fully understand what the humpback whales are doing, we need to know many more details. Their brains are larger than ours and equally complicated, according to most of the criteria used by neuroanatomists. The cerebral cortex is just as convoluted, and the differentiation of brain cell layers is as distinct as in human brains. While we know far too little to speculate about what the hump-

back whales are thinking, it seems likely that they are anticipating
the taste of food and the sensation of swallowing something more
than sea water.

We must turn to species whose feeding tactics have been studied
in much greater detail. In recent years many theoretical ecologists
have analyzed and evaluated what are called optimal foraging
theories, which predict how animals should allocate their time
and effort to obtain the most food with the least exertion. These
theoreticians are fond of mathematical equations predicting how
animals ought to behave in order to operate at maximum efficiency. Underlying this approach is the confident belief that natural selection has influenced not only the animals' structure and
functioning but also their behavior. As Darwin first convinced
the world, the more efficiently an animal is constructed, and the
more efficiently it behaves, the more offspring it will contribute
to future generations. This idea forms the underpinning of almost
all thinking, observing, and experimenting concerned with feeding
behavior. This basic theorem is almost a truism, and the whole
field of evolutionary biology has been criticized by some philosophers as an elaborate rehashing of the obvious. Inept animals
are clearly less likely to survive and reproduce; therefore it is not
surprising that those that have survived are reasonably efficient.

We need to consider several subtleties, however, to achieve a
balanced appreciation of evolutionary biology in general, and
foraging behavior in particular. First of all, evolutionary biologists
assume that the behavior has a significant hereditary component.
If the feeding behavior of the offspring did not closely resemble
that of the parents, the evolutionary advantage of any improvement in foraging would not last for many generations. For this
kind of basic theorizing it does not matter just *how* the behavior
of offspring comes to resemble that of parents; the DNA of the
animal's genes might directly control the feeding behavior, or the
offspring might simply learn by imitating their parents, with no
genetic influence. Both genetic and environmental factors doubtless influence behavior, and complex interactions of the two categories are also important, but very difficult to analyze. Evolutionary
theorists tend not to worry about these complexities; they simply
assume some correlation in feeding or other behavior between
generations, and that efficient feeders will, through natural selection, increase in numbers relative to those that use their time
and energy less productively.

Still other complications are usually ignored by scientific investigators, although they are likely to occur to thoughtful people less directly involved in behavioral ecology. If natural selection is so powerful, why have all wild animals not reached a perfect state of behavioral adaptation? Since this seems not to be the case, evolutionary biologists have been forced to postulate other considerations that stand in the way of perfect adaptation. One of the simplest is the recognition that natural environments, particularly food supplies, are by no means constant over space and time. If animals evolved to be perfectly adapted to one ecological situation, they might find themselves in different circumstances that would call for a quite different set of adaptations. Nor are the patterns of variation in food supplies predictable from time to time and place to place. Thus theorists recognize that the best practicable adaptation is not an absolutely constant and rigid set of behavior patterns. It is more efficient to be able to change one's feeding tactics according to the available food and the problems of securing it.

The behavioral ecologists who study such questions tend to divide into two groups: the theorists who feel most at home with mathematical formulations and tend to make simplifying assumptions so their equations will be mathematically manageable, and the field naturalists who study whether animals actually behave according to the theories their colleagues have developed. Because the real world is so complex, and because continuous and detailed observations of wild animals are often formidably difficult, there is an intermediate group that studies captive animals whose food supplies have been arranged in conveniently analyzed patterns. These patterns may include variations in the abundance of food, whether it is conspicuous or concealed, whether it is concentrated into clumps or widely dispersed, and so forth.

All of these approaches have yielded information upon which we can draw in considering what foraging animals may be thinking about. It is important to recognize that the scientists who gathered the pertinent data were not interested in my cognitive questions. They have operated almost universally under the typical mid-twentieth-century opinion that relegates questions of subjective mental experience to the limbo of the nonscientific. For the most part, they were simply trying to measure how well the actual behavior of feeding animals agreed with various theoretical predictions.

Searching Images

An especially important aspect of food gathering is the concentration of searching effort on things that may be edible. If foraging animals scrutinized every detail of their environments with equal attention, they would waste an enormous amount of time on unimportant items. Instead, each foraging creature searches for those things that look, sound, or smell like possible food. Some animals use other sensory systems, such as certain sharks that are sensitive to electric currents that permit them to detect muscle action potentials from hidden prey. But because visual clues are much easier for us to analyze than odors or even sounds, most of the studies of food searching have involved vision.

Several careful studies of foraging birds have strongly indicated that they look for a particular pattern that tells them where food can be found. Learned patterns may include the barely perceptible outline of a cryptically colored moth resting on the bark of a tree trunk or some marking provided by human experimenters to show the animal where it can find food. Somewhere in the animal's brain there must be a mechanism for recognizing what is called a searching image.

One of the most thorough studies was carried out by Harvey Croze (1970) as a graduate student at Oxford University. He inquired how the carrion crows of England learned to locate food marked by a novel visual signal. Croze laid out a row of empty half mussel shells on a sandy beach near where the crows were accustomed to gather mussels at low tide. Beside each of the empty shells, located on open sand, convex side up, where they were clearly visible, Croze placed small pieces of beef. About five hours later the birds had taken every piece of meat. On the following day he laid out twenty-five mussel shells with pieces of meat hidden underneath each one. Within a very short time the crows returned, turned over twenty-three of the twenty-five shells, and ate the meat. They thus had learned quickly that mussel shells lying on the sand, which normally would be empty, had suddenly become sources of a new, tasty food. On the following day Croze again laid out twenty-five shells but this time the meat was buried in the sand underneath each one. Soon after daylight the crows returned and, after turning the shells over and finding nothing edible, dug with their bills in the sand until they found the meat. One cannot rule out the possibility that odors played a role in this behavior; although olfaction is not well developed in birds, pigeons can learn to discriminate between odors (Hutton et al., 1974). So the crows may have located the buried meat in part by its odor.

Observations and experiments make it clear that crows and similar birds, such as blue jays, are quite expert at watching for new kinds of objects that are edible, or that mark the location of accessible food. Such learning operates in both directions. When Croze stopped placing meat under the mussel shells, the crows continued for a time to turn the shells over, but gradually paid less and less attention to them. Yet when they were turning mussel shells only occasionally, if they found a single one that was baited, they would turn over many more than they had immediately before. This corresponds roughly to many situations in nature when some natural object may sometimes indicate the presence of food. If it does so on one occasion, it pays for the animal to inspect similar objects for a short time afterward.

These and similar experiments have led ethologists to conclude that many animals use searching images when they forage. Something in the animal's central nervous system causes them to recognize a certain pattern and approach, explore, and make other food-searching motions. Searching images may be quite specific, as when the crows learned that mussel shells on a beach had suddenly become markers of food, or they may be more diffuse and general, as when animals learn that a particular type of habitat is where some type of food is most likely to be found. When they have just recognized a new searching image, crows and other animals may consciously think about their fruitful discovery.

Feeding Behavior of Marsh-Nesting Blackbirds

The behavioral ecology of redwinged blackbirds and the closely related yellowheaded blackbirds has been studied in detail by Gordon Orians (1980) of the University of Washington. These two species are abundant in marshes in the northwestern United States and western Canada, where they can be observed throughout a large fraction of their daily activities. Orians and his colleagues concentrated on the season when the blackbirds were raising young, a time when nesting birds are under great pressure to obtain enough food to feed their nestlings. Furthermore, in this situation evolutionary selection operates directly. The amount and quality of food provided to the growing young make a large and measurable difference in the number that survive and leave the nest as healthy fledglings.

In the area studied, both species nest in vegetation growing in shallow water. The redwing is a strongly territorial bird; in the spring the males arrive first in the breeding area, where they establish and defend an area of marsh or adjacent upland. Somewhat later the females arrive and a number, sometimes as many as a dozen, construct their nests within one male's territory. The females do all of the nest building and incubation and nearly all of the feeding of nestlings, although after the young have left the nests, the males also feed them. The larger yellowheaded blackbird has similar habits except that the males do help feed the nestlings. In each species a territory-holding male mates with several females. Many other adult males without territories are also present in the breeding area but contribute little or nothing to the breeding.

Both species feed extensively on adult aquatic insects that have just emerged from the water; the redwings also feed on insects from the dry upland areas. The yellowheads, which are larger, exclude redwings not only from their nesting territories but from the richest sources of aquatic insects. Marshes vary widely in the abundance of insect food, and some are inhabited by numerous blackbirds, while others are scarcely used at all. In Orians' study area two major factors in the abundance of insects were changes in the density of vegetation and the invasion of lakes or swamps by carp that ate most of the immature aquatic insects before they could emerge and become prey for blackbirds.

One of the earliest choices made by the male blackbird in each breeding season is the selection of an actual territory. Out of extensive areas of marsh they choose one small region where they sing and from which they attempt to exclude other males by threats and sometimes actual fighting. When the females arrive a week or two later, they visit several territories before choosing one in which to mate and build their nests. How do the males and females make these choices? The territory must be selected well before the abundance of insect food is critically important. In fact, at that time hardly any aquatic insects have emerged, so the choice of breeding territory must be guided by some other factor.

One could suppose that the blackbirds make no choice at all, that after the first males choose randomly, others are attracted to adjacent territories. But the careful studies by Orians and his colleagues showed that out of the extensive areas of available marsh, the blackbirds selected those that later produced the richest harvest of aquatic and terrestrial insects. In terms of food supplies, the blackbirds approximated what the theorists have called an ideal free distribution, meaning simply that the density of the bird population corresponds to the quality of habitat. Where food and other requirements are most abundant there are just enough more birds to utilize these resources with maximum efficiency. In an ideal free distribution, any bird that moved to a different area would find inferior conditions, less food, greater effort required to gather a given amount of food, or all three. Of course many factors also enter into the suitability of a territory, such as appropriate vegetation for nesting and relative safety from predators.

This nearly ideal distribution stimulates the behavioral ecologist to ask how the male blackbird decides where to establish a territory and how the later-arriving females decide where to mate and build nests. Perhaps they are guided by memory and tradition. If the marshy areas remain the same for long enough, birds might simply remember where they nested last year or where their parents raised them. Unfortunately for this theory, however, the marshes in the Pacific Northwest change rapidly from year to year as a result of variations in water level, invasion of lakes by carp, and other ecological changes.

These birds must make their choices each year on the basis of some properties of the marshes at the time they first arrive. Newly arrived birds often forage at the air-water interface when they are selecting their territories. For hours on end the females seem to ignore the vigorous displays of the males and instead spend a great deal of time at the edge of the water. Most of the aquatic insects are still larval or nymphal stages swimming about underwater where the blackbirds cannot reach them, although they might see them if they look closely. Conceivably the birds might look for aquatic insect larvae, but Orians interprets his observations more conservatively:

It seems likely that the patterns of open water and emergent vegetation may be utilized, since marshes that will later provide abundant emerging insects have certain properties. For example the density of stalks of aquatic vegetation is important. If they are too closely spaced, relatively few insect larvae will be present, but many will emerge at the outer edge of this vegetation. This almost certainly improves foraging at the outer edge of vegetation but makes it less productive elsewhere. Of the two species the larger yellowheads tend to occupy the more open vegetation, while the redwings nest in denser vegetation, usually closer to the shore, where food is less abundant. This may result from the actual territorial exclusion of redwings by the larger yellowheads, but it may also reflect the fact that the redwings do more of their foraging on upland areas.

It is also interesting that the yellowheads tended to nest in marshy areas where no continuous stand of trees extended up more than about 30 degrees from the horizon. Evidence that this was an important factor in territory selection was provided by a particular pond where tall cliffs rose abruptly from the water's edge. These cliffs did not prevent the formation of ideal foraging conditions, but the yellowheads did not nest there, suggesting that the presence of any large continuous obstruction rising high above the horizon was a factor in the choice of territory. Perhaps the cliffs provide convenient places for hawks to roost or build their nests. Unfortunately this "natural experiment" occurred in only one instance, so one cannot rely heavily on the resulting inference. But I mention this as an example of the factors that seem to be important in selection of nesting territories.

What is it like to *be* a male blackbird arriving after a long migration and deciding where to establish a territory? How do you, as a female blackbird, choose among the many territory-holding males? Behavioral ecologists tend to assume that some genetically fixed behavior pattern dictates the choices. They seldom consider whether the birds ask themselves "Will there be lots of insects here?" But there is no reason why simple thinking about food or danger should not be an important attribute forged by natural selection. When one considers the variety of choices open to these birds, and the critical importance of future food supplies, it is possible that their behavior is influenced by simple thoughts about the situation.

The actual capture of food is more difficult to study, because the insect prey are small, and in many cases the blackbirds catching insects are too far away for close inspection, even with the aid of binoculars. Orians and his colleagues used several methods to learn what quantities of different types of insects were taken. One method was to put around the neck of a nestling a loose pipecleaner collar. This was not tight enough to prevent breathing, but it did prevent the nestling from swallowing the insects brought by the parent. The accumulation of unswallowed insects could then be removed for study. This procedure showed that as many as ten insects were delivered at a time.

By careful observation and long practice, Orians and his colleagues also learned to identify through binoculars many insects as they were captured. Sometimes a bird with a large load will drop what it has while trying to capture an additional insect, but in that case it will always pick up the previously gathered prey and then carry the whole lot back to the nest. Theories about optimal foraging predict that a bird gathering food close to the nest should return more often with smaller loads, since the return trip requires less time and effort, and the young can be digesting several smaller feedings. This prediction was at least roughly confirmed by the observations.

The busy mothers gathering food for their nestlings must also feed themselves. Blackbirds usually swallowed the first few insects captured on any one foraging trip before beginning to gather food for the young. What do such parent birds think or feel during this very energetic sort of behavior, which occupies most of their waking hours? Why do they feed nestlings at all? The obvious evolutionary answer is that feeding the young is necessary for reproduction; birds with this type of behavior are successful in reproducing, whereas those lacking it would leave no descendants at all. But what is it like to be an incubating bird on the day when those warm, smooth eggs crack open and unleash a number of damp, squirming, noisy creatures with gaping mouths? Something clearly motivates the parent bird to gather food and insert it into those mouths. But what does the parent bird experience? If this is the bird's first brood, its only experience with the situation is its own nestling stage a year earlier, which it might remember. Do the parents think, "These young need food" or "If I put food in their mouths they will stop squawking"?

Another interesting observation made by Orians was that when they are not feeding young, blackbirds often eat dragonflies; however, if they capture a dragonfly when they are gathering food for nestlings, they immediately take it back to the nest. In the area of these studies, dragonflies are quite large and provide excellent nutrition for the young. The parents themselves never seem to eat them when they are caring for nestlings. Something causes their behavior with regard to dragonflies to be drastically different when they are feeding only themselves and the special and demanding times when they are feeding young. Do they think something like, "This big insect will feed those babies nicely. I will take it back to them and catch something else for myself?"

Prey Selection by Wagtails

Among the best studies of prey selection and decisions about how to hunt for food are those of M. B. Davies on Port Meadows along the banks of the Isis River in Oxford, England (Davies, 1977; Krebs and Davies, 1978). Davies studied pied wagtails, which are year-round residents of southern England, and a closely related species, the yellow wagtail, which is a summer visitor. His observations were made early in the spring before the birds started to breed. They were gathering food only for themselves, although they may have been putting on weight in preparation for the nesting season that would follow within a few weeks. It was quite easy to observe the wagtails on Port Meadows partly because the grass was heavily grazed by cows all year, and the numerous dung pats provided a major source of food for the insects on which the wagtails fed. Every day when one of these birds set out to feed, it had to decide where to fly, whether to join a flock or hunt by itself, what insects to select, and how to catch them.

Davies found that the principal food of these two species of wagtails in spring were flies that lay their eggs in dung pats and a variety of small insects living in shallow water at the edge of the river. One bird at a time would hunt insects at a dung pat, but at the water's edge a large group usually fed simultaneously at the same area. After the most easily available insects have been captured, the bird must decide whether to continue searching for more food at the partly depleted area or move elsewhere. If it decides to move, it must choose a spot on the bank or go to a particular dung pat. The birds probably learn where food is likely to be available. For example, the fly pupae in the dung pats emerge as edible adult flies in a gradual sequence, so where there were several yesterday there are likely to be others today. Conversely, where there were few or none, it is less likely that new flies will emerge today.

Even in this relatively simple foraging situation, the bird must weigh many factors to make the most of its time and effort. Optimization is vitally important, because even in the spring before they are feeding young (but when insects are also less abundant than later in the season) these wagtails spent about 90 percent of their daylight hours gathering food. Any appreciable reduction in foraging efficiency would result in weight loss, poor health, and probable failure to raise a full complement of young. The selective pressure to make optimal feeding choices is very strong.

In general the Oxford wagtails shifted from dung pats to water's edge and vice versa efficiently, so that they obtained close to the maximum amount of food with minimum expenditure of time and effort. They concentrated their efforts where food was plentiful, and when it became depleted they moved on, not at random but to areas that were reasonably productive. When we see people behaving efficiently, we assume they have given some thought to what they are doing. Can we draw a similar inference about wagtails? Scientists who prefer a reductionist view of life argue that efficiency of a mechanism is no evidence of conscious awareness. But foraging optimally requires accurate evaluation of so many aspects of a changing environment that conscious thinking about the situation may be the most effective procedure.

Dropping Shellfish on Hard Surfaces

Herring gulls have often been observed to carry shellfish to rocky areas or hard pavements, then drop them, sometimes repeatedly, until the shells break and the gull can eat the mollusk. In some areas crows also feed in this way, and both species have been studied with considerable care by ethologists. This behavior is quite different from simply discovering a morsel of food, seizing it, breaking it up, and swallowing the pieces.

Especially thorough studies of shell dropping by herring gulls have been carried out by Benjamin Beck (1980, 1982). The gulls pick up clams, whelks, or shells inhabited by hermit crabs at low tide and carry them for considerable distances, typically 30 to 200 meters, to places where they can drop and break the shells. Although ethologists have reported that gulls in some areas are not selective about where they drop shells, in Beck's study area and many other places the gulls do choose rocky areas, roads, paved parking lots, or sea walls. The herring gulls observed by Beck flew quite low and could not see the hard surface they were flying toward; they had evidently learned where to go for this purpose. An indication of the selectivity of Beck's gulls is that of the area over which one group was accustomed to carry their shells, only about 1 percent was a rocky sea wall where the shells would break consistently; but 90 percent of the drops were directed at this wall. In another area, a large parking lot, the gulls dropped their shells from a greater height, perhaps because they knew there was no danger of missing the target. Shell dropping is probably learned by observation of other older gulls. Young herring gulls were less efficient than adults; they dropped very few shells, but they often dropped other objects in play.

Detailed studies of shell breaking by crows were carried out by R. Zach (1978) on an island near the coast of British Columbia. A small breeding population of crows commonly gathered food at low tide, including whelks, which have quite a heavy shell two to five centimeters long. Each pair of crows had a special section of the beach for foraging and a particular rocky area where they dropped whelks. The selection of dropping sites was important, for the rocks had to be away from the shore so that the whelks would not bounce into the water, and fairly level so that edible fragments would not fall into deep crevices.

If the size of the whelks varied, the crows selected the larger ones, and they rejected shells that did not contain a normal live mollusk. When one observes such preferences under natural conditions, it is very difficult to know what factors are influencing the bird's behavior, but Zach clarified the matter by means of some simple experiments. He stuffed empty whelk shells with either very light material or with material of similar density to the living mollusk, then glued back in place the horny operculum that closes the shell. He found that the crows selected shells of normal weight or density, rejecting most of those that were much lighter. Light-weight shells ordinarily would mean a dead and dried-up corpse inside rather than a fresh and edible mollusk.

Zach also studied the shell-dropping tactics in some detail. The crows dropped whelks from heights between approximately 3 and 8 meters. The chances appeared to be about one in four that a whelk would break when dropped, so many of them had to be dropped repeatedly. Unless they were disturbed, the crows would persist until the shell broke, which could take as many as twenty drops. It would seem more efficient for the crows to fly higher, so that the shells would be more likely to break, but the flight patterns suggested that the birds had some difficulty seeing where the whelk had fallen. By dropping the whelk as they descended from their maximum altitude, they could watch its trajectory, and they were less likely to lose it. Whelks dropped from a greater height also seemed more likely to shatter, making the soft parts more difficult to retrieve and perhaps less pleasant to eat because of small bits of broken shell. Sometimes the crows dipped broken whelks into fresh water puddles before eating them, apparently to remove fragments of shell.

Zach's main interest was to determine whether the foraging

behavior was optimal, so that the crows gained as much food energy for a given effort as the situation permitted. Like the Oxford wagtails, they did very nearly as well as was possible by choosing only large whelks and only those of approximately the right weight, by dropping the shells only on suitable rocky areas, by choosing the best height for dropping, and by persevering with each whelk until it broke rather than giving up and searching for another one.

Since only a small fraction of crows collect whelks and drop them on rocks, this type of feeding had probably developed quite recently. Many crows on the island Zach studied were feeding in this way, however, so the habit had probably spread throughout a small local population, presumably by observational learning. No one observed the first crow's discovery of this type of feeding nor its adoption by other crows, but it must have required a certain amount of simple but creative thinking by the birds that first began dropping shells. It is interesting that one individual crow was observed to depart from custom by carrying and dropping two whelks simultaneously. Such individually enterprising behavior is reported from time to time, but ethologists have not emphasized that this suggests thinking.

Foraging Decisions by Bumblebees

One might not expect a high order of thoughtful decision making from bumblebees, but if we study their behavior closely, we find that gathering nectar from flowers is a complex and efficient process. G. H. Pyke (1979) studied a particular species of bumblebee feeding on monkshood flowers, which consist of large clusters, in the Colorado mountains. The flowers vary considerably in the amount of nectar they contain, depending in part on whether an insect has already visited the flower and sucked out most of the nectar. Pyke studied individual marked bumblebees while they were visiting flowers that had not been visited by other insects. Even if all the flowers on an inflorescence held a good supply of nectar, a bumblebee had to visit several flowers to fill her stomach. They almost always visited first the lowest or next to lowest flower, then moved upward, usually selecting the closest flower not already visited. In only 4 out of 482 observations did a bumblebee return to the same flower twice. This suggests that a bumblebee can remember at least for a short time which flower she has previously visited, or that perhaps she leaves a scent mark and simply avoids flowers with this odor. But even when a bumblebee has taken nectar from two or three other flowers in the meantime, she rarely flies back near one she has already visited. Pyke summarized these observations by stating, in terms as simple as the data allowed, that the bumblebees follow rules of movement, such as: "Start at the lowest flower on a given inflorescence, then move to the closest flower not just visited, unless the last movement had been downward and was not in fact the first switch from one flower to another on a particular inflorescence. In the latter case, move to the closest higher flower not just visited." It seems clear that no other pattern would have been appreciably more efficient, according to the available data.

If all this sounds a bit complicated for the central nervous system of a bumblebee, we should ask whether we have underestimated the capabilities of these animals. Pyke's study is only one example showing that even animals we think of as quite simple behave according to ingeniously effective rules.

Vampire Finches

Although the finches on the various Galapagos Islands are closely related to each other, they apparently do not fly from one island to another. As a result of isolation for several thousand years, local races or subspecies have evolved. Darwin's recognition that species are not fixed and immutable was stimulated to a considerable extent by his pioneering studies of these island-to-island differences. For this reason the whole set of local races is often known as Darwin's finches. All are versatile birds, which sometimes use thorns or twigs as simple tools, to search for insects. In the course of studying the many fascinating biological phenomena presented by animals living on the Galapagos Islands, Bowman and Billeb (1965) discovered that one local race of Darwin's finch has learned to feed on the blood of larger birds.

On Wenman Island, which is about 100 km from other islands in the archipelago, the Darwin's finches have learned to feed on the blood of large boobies. This habit probably began as a mutually advantageous association in which the finches removed the large parasitic insects that infest the boobies when they are nesting. In several other parts of the world small birds have developed the habit of feeding on parasites that they pick off the skin of much larger animals. On Wenman Island, however, the finches have learned to peck vigorously at the base of the boobies' feathers, most often near the elbow of the folded wing, drawing enough blood to drink. The boobies often try to dislodge the finches, but the finches usually succeed in the end. The blood-feeding habit is well established and enables the finches to survive and reproduce on a barren island where other food is very scarce for long periods. The boobies also nest on other islands, where very similar finches could feed in the same way, but they do not. The fact that other finches fail to exploit an available food resource demonstrates that evolutionary adaptation is not perfect.

How did these finches become "vampires"? Darwin's finches search widely for insects, often poking into crevices to find them. The change from searching for insects in vegetation, pebbles, or other nonliving surfaces to taking parasites from the feathers of a large stationary bird is not a very great one, and presumably the boobies tolerated these small birds since there was some benefit in having the parasites removed. Some of the bird lice and louse flies are quite large and conspicuous when they come to the surface of the all-white feathers. At other times, however, these ectoparasites burrow down into the feathers, and a finch poking into the feathers in pursuit could quite easily peck the booby's skin vigorously enough to cause bleeding. At some time, probably not too long ago, one or more enterprising finches must have taken the significant further step of discovering that the blood was edible and nourishing, perhaps when the finch missed the parasite it was pecking at. The behavior spread, probably through observational learning, and became a normal part of the finches' feeding behavior.

Can we infer any sort of thinking or mental experience in such a situation? E. C. Tolman (1932, 1966) observed that rats trained to carry out a complicated pattern of behavior to obtain food act as though surprised if the food reward they expect is suddenly not available. Other psychologists have studied this sort of expectancy and surprise, and there is little doubt that when an animal might reasonably expect a certain outcome it does act surprised and disappointed when this is not forthcoming. In the case of the pioneer vampire finches discovering a new source of food, the birds would have been surprised to find that tasty and nutritious blood, previously obtained by swallowing parasites, could also be gathered by pecking at a booby's skin. We can only speculate about these events, but the striking nature of the change, and the enterprising nature of the feeding behavior of Galapagos finches in other circumstances, suggest that a simple form of insightful thinking may have occurred when they first discovered this new source of food. If a Wenman finch is blown to another island, will the native finches watch it exploit the boobies in this fashion, and will observational learning then lead to a new population of vampire finches?

Laboratory animals also learn to exploit novel sources of food and to use patterns of behavior quite different from their normal ways of feeding (Mackintosh, 1974). In many cases the animal must spend some time learning that food will be made available at an unfamiliar place after it has performed certain actions, such as pressing a bar or pecking into a newly opened cavity. Laboratory situations could easily be set up to approximate the experience of a Galapagos finch suddenly discovering that pecking harder at a booby's skin yields a new type of food. Perhaps there is nothing so very special about the Galapagos vampire finches; perhaps even White Carneaux pigeons experience the same sort of surprise as they learn that pecking a lighted circle of glass causes a metallic cavity to open and bring food within reach. Strict behaviorists argue that in both cases it is silly to postulate that thoughts accompany the learning, but it seems to me that thinking is more likely in these situations than when animals are repeating a standardized pattern of feeding behavior many times over. This was Tolman's point, which led him to suggest that rats were consciously aware of some aspects of learning a new behavior to obtain food, water, or some other reward.

Adaptability in Insect Hunting

In western Europe there are several relatives of the North American chickadees and titmice, those acrobatic and entertaining visitors to thousands of bird feeders. These birds, called tits in England, all belong to the same genus (*Parus*). Because the birds do well in captivity, J. R. Krebs and others have studied the nature and efficiency of their feeding tactics when they search for insect prey. The purpose of these experiments was to test mathematical theories about optimal foraging behavior; Krebs and his colleagues did not admit that they were interested in any thoughts or feelings that may have accompanied the foraging behavior.

In one set of experiments Krebs, MacRoberts, and Cullen (1972), Krebs and Davies (1978), and Krebs (1979) studied how great tits (*Parus major*) coped with the challenging problems of foraging for concealed food. To standardize conditions so they could make meaningful quantitative comparisons of different birds or of the same birds under different conditions, Krebs and his colleagues did not use the usual insect prey hidden on natural vegetation. Instead they used beetle larvae, called mealworms, which are approximately 3 mm in diameter and about 25 mm long. Many insectivorous animals eat them avidly in captivity.

Krebs and his colleagues were interested both in the way captive great tits would learn to find hidden mealworms in different situations and how several birds hungrily foraging in the same cage would interact. The scientists used several types of hiding places, including plastic cups, called "hoppies"; pingpong balls cut in half, called "pingies"; and small blocks of wood with a narrow hole drilled in the top, designated "milkies." All of the containers were filled with sawdust or bits of paper and the openings covered with masking tape; only a fraction contained mealworms. The birds were first given unburied mealworms, then containers with mealworms buried in sawdust but not covered with tape, and then the covered containers. They learned surprisingly quickly that they could sometimes find food by pecking through and tearing off the tape.

"Barkies," a fourth type of hiding place, consisted of small strips of masking tape stuck on the trunks of trees in the cage. The mealworm hidden underneath the tape showed as only a small bump. To make the problem more difficult, some pieces of tape concealed only a short piece of thick string. The birds never learned to discriminate between the two kinds of bump and were equally likely to attack hidden string or hidden mealworms. All of these elaborate arrangements were designed to simulate the task an insectivorous bird faces when searching large areas of natural vegetation for those few spots where something edible can perhaps be uncovered by probing or pulling off loose bark.

The results of these experiments showed that when one great tit discovered that a particular type of container sometimes contained mealworms, it would search for similar containers, including tape stuck against the tree bark. More important, if not altogether surprising, was that the other birds in the cage also began to look for similar hiding places. In other words, these birds learned a great deal about where to locate food by observing where their companions were finding it. Individuals differed in their searching tactics. In general the dominant bird chased others away from food and took what it wanted, and it visited more containers per minute than the other birds.

Individual birds also specialized in particular feeding methods. Some concentrated on the hoppies, turning over the pieces of paper that sometimes concealed a mealworm and peering for several seconds into the container. Others wallowed in the container and threw out the pieces of paper, exposing any hidden mealworms. One discovered that by pecking through the tape he could peer into the container and see whether a mealworm was there. Some opened the milkies by hammering through the tape, while others would pull away one edge of the tape. All of these individual patterns make it clear that the birds were actively—and perhaps consciously—trying to find the concealed food. Their foraging behavior was certainly not rigid and stereotyped; it changed rapidly according to the circumstances and the results obtained not only by the individual but by its companions.

The behavior of these birds reminds us of the Japanese macaques that learned to wash potatoes provided under seminatural conditions and also learned to separate grain from inedible material by throwing it into the water; the kernels of grain would float while the inorganic sand and other particles tended to sink (Kawai, 1965). These new types of food handling were first devised by a few monkeys, then were gradually acquired by other members of their social group through observational learning.

Behavioral versatility came into play in a spectacular fasion in the 1930s when two species of tits discovered that milk bottles delivered to British doorsteps could be a source of food (Fisher and Hinde, 1949; Hinde and Fisher, 1951). At that time milk bottle tops were made of soft metal foil, and the milk was not homogenized, so the cream rose to the top of the bottle. One or more birds discovered that the same type of behavior used to get at insects hidden under tree bark could also be used to get cream from milk bottles. The people whose milk was disturbed immediately noticed it, and careful studies were made of the gradual spread of this behavior throughout much of England. A change in the technology of covering milk bottles eventually ended the whole business, but meanwhile thousands of birds had learned, almost certainly through observation, to exploit a newly available food source.

The discovery of milk bottles by great tits is seldom emphasized in contemporary textbooks and monographs on animal behavior. Ethologists and behavioral ecologists seem to lose interest when animals do something suggestive of versatile thinking. In the hundreds of pages devoted to more mundane aspects of behavior, this outstanding example of unexpectedly enterprising learning does not seem especially significant to contemporary ethologists who have little interest in animal behavior that seems to require inventive thinking.

Oystercatchers Feeding on Mussels

The oystercatcher is a large shore bird related to the sandpipers, with a large, conspicuous red bill. Much of the time oystercatchers feed on mollusks exposed at low tide, but they sometimes turn to other prey, such as earthworms exposed in plowed fields. The English behavioral ecologist Norton-Griffiths (1967, 1969) discovered by patient observation how these birds deal with tough mussel shells. At low tide some mussels are fully exposed and others are covered by shallow water. When they are fully exposed, the oystercatchers seize them in their bills, pull them loose from the substrate, and carry them to a patch of hard sand. There the bird turns the mussel shell so that its flat ventral surface is uppermost. Even though this is not the most stable position, the bird keeps the mussel so oriented as it hammers open the shell. To study what it takes to hammer open these shells, Norton-Griffiths built a mussel-cracking machine, using a close copy of an oystercatcher bill as the pick. The flat ventral surface of the shell proved to be the most easily broken part, indicating that the oystercatchers had learned how to open a mussel in the easiest way. Each bird also learned where the sand was suitably hard and brought numerous mussels to the same spot.

When the mussels were covered by shallow water, the oystercatchers opened them in an entirely different way. They searched for slightly opened shells into which the bird would stab its bill, cutting through the abductor muscle that closes the shell. Then it would carry the shell to some convenient location for eating.

Norton-Griffiths, like any sensible person observing what I have just described, assumed at first that the oystercatchers adapted their mussel-opening techniques to varying situations. But when he marked individuals and watched their feeding behavior, he found that each one specialized in one or the other type of feeding. Further observations strongly indicated that the young learned this when they began feeding with their parents, by watching them locate and open mussels in one of the two ways described above. Each oystercatcher learned one of these techniques, rather than the behavior being rigidly predetermined by genetic influences.

What is it like to be an oystercatcher feeding in this somewhat complicated fashion? And what is it like to be a young oyster-catcher learning how to open mussels? The fact that there are two radically different techniques, and that each bird seems to learn one but not the other by imitating its parents, suggests a certain amount of thoughtful attention and versatile adaptability. Presumably, adults with extensive experience become relatively stereotyped in their mussel-opening behavior, but at the time youngsters learn this behavior, going through imperfect inter-mediate stages, the process was probably challenging enough to require some thought.

Food Caches

Many animals, even laboratory rats, store excess food when it is abundant. Under natural conditions much of the stored food is recovered and eaten later, often weeks or months later. Animals often go to some trouble to select sheltered places for their food caches and to cover them or take other steps to keep other animals from getting their food. Squirrels that hide seeds and nuts are familiar examples, and they recover a large fraction of those they store. But it is not clear just how the animal relocates the stored food. Scientists, skeptical that squirrels could remember where they stored food, have suggested that they might detect it by odor, or by seeing signs of disturbance to the ground where the food is buried, or perhaps by a wasteful sort of random searching. As in other instances where complex animal behavior suggests con-scious thinking, ethologists have been slow to turn their attention to this potentially significant subject.

One recent and significant exception is the work of Russell P. Balda and his colleagues at Northern Arizona University, who have carefully investigated the large-scale seed caching of the Clark's nutcracker, a relative of the crows and jays, which lives in alpine environments of western North America where food is scarce during long cold winters (Balda, 1980; Vander Wall and Balda, 1981; Vander Wall, 1982). During the autumn the nut-crackers spend a great deal of time secreting pine seeds in crevices or burying them in the ground. In a year when seeds are plentiful a single bird may hide as many as 33,000 pine seeds. Each cache usually contains from two to five seeds. Conservative estimates indicate that to survive through a typical winter a Clark's nutcracker must relocate approximately a thousand of the caches it made the previous autumn.

In laboratory experiments a nutcracker was induced to bury pine seeds in sand spread over the floor of a cage. After being kept away from this cage for a month it recovered and consumed a substantial fraction of the seeds—far more than could be explained by random searching. Meanwhile the experimenters had buried numerous other seeds of the same kind in the sandy floor. The bird found very few of these, indicating that it was not searching at random but returning to the spots where it had buried its own seeds. If the bird was locating the seeds by odor, presumably it would have recovered many of those buried by the scientists. In more definitive experiments, conspicuous logs and stones were placed on the floor of the cage, and the bird buried most of its seeds near these. Its success at relocating seeds was greater when such landmarks were available. But when the landmarks were shifted during the month when the bird was kept out of the cage, on its return it would search near a certain part of a log, as if it remembered the location of each cache in relation to the log; its recovery rate was very low under these conditions.

The quantitative aspects of this behavior are truly impressive, especially when we recall the changes in the natural environment between the time when seeds are hidden and the wintery conditions when many of them are retrieved. By winter the vegetation has changed and the ground is often covered with snow. Would it be helpful for the nutcracker to think about where it had hidden its food? One cannot say with certainty, but the whole behavior pattern certainly is suggestive of thought, especially when we recall that when the bird buries seeds it smoothes the ground and tries to remove any signs of disturbance. At the very least we must postulate not just a single searching image, but hundreds of them. Such a prodigious number of specific visual memories of particular stumps and rocks boggles our minds. Perhaps some mental map or patterned scheme helps these birds. After burying their seeds, they presumably spend much time flying over the area where their caches are located, and perhaps they maintain a map-like memory of those features that are important to them. Food storage points might be more easily recalled as parts of such a pattern than as hundreds of individual features.

The traditional way to think about such behavior constrains us to imagine a series of separate instinctive actions, first picking up seeds, then carrying them to a hiding place, burying them, and covering them in a manner that makes the hole less conspicuous. All this is supposed to occur without the bird having the slightest thought of returning and eating the seeds. Then, weeks or months later, when the animal is hungry it remembers hiding food at a certain place; it returns there, retrieves the food, and eats it.

Further experiments of this type have been carried out with marsh tits in England by Cowie, Krebs, and Sherry (1981) and by Sherry, Krebs, and Cowie (1981); the whole subject has been thoughtfully reviewed by Shettleworth (1983). These small relatives of North American chickadees, and probably other birds as well, not only hide large numbers of seeds, but while doing so look closely at the hiding places, suggesting that they are thinking about returning later to find the food they are hiding. We cannot reach any more definite conclusions, but these experiments have certainly shown the richness of impressive accomplishment in the natural behavior of animals when ethologists examine it carefully.

This has been a rapid excursion through the ecology of feeding behavior, rather like an eight-day packaged tour of a wholly unfamiliar continent. So much is different, and the contacts with the inhabitants are so limited, yet from the blurred images a thoughtful and sympathetic traveler may begin to understand what it is like to be one of these other people. Of course, much deeper and more accurate understanding can be reached by a devoted anthropologist who settles in for a long time in the society, learning the language, eating the local foods, participating in festivities and rituals, and even sharing some of the local afflictions.

In much the same way, though to a much more limited degree, cognitive ethologists seek to learn what the lives of other species are really like. In this guided tour of some high points in cognitive ethology—the seven most famous temples in our four-hour dash through an ancient city—one should try hard to avoid the simplistic fallacy that humpback whales, crows, bumblebees, or even earthworms do *only* those things I have mentioned. My examples of feeding behavior were selected primarily because they have been well studied. Both the animals and their behavior had to be easily observable, so we could be confident that the behavior was truly characteristic. This is why so many of the examples were diurnal birds. Most mammals are active at night, fishes and whales do most of their hunting under water, and insects rely primarily on odors we cannot sense, all of which make it very difficult to gather detailed information on their feeding behavior.

The ethologists who carried out these laborious investigations have not stated that they were studying what it is like to be a crow, an earthworm, or an oystercatcher. In spite of this restriction of outlook, they have discovered many kinds of ingeniously adapted behavior in various animals. We can hope that in the future, behavioral ecology may reveal much more about what animals think or feel as they cope with the problems of finding food.

4 Predators and Prey

Catching mobile and elusive prey calls for flexible behavior that can be rapidly and effectively adapted to changing circumstances. The stakes are extremely high. For the prey it is literally a matter of life and death. For the predator, success or failure in a particular effort is less crucial, but its survival and reproduction depend on succeeding reasonably often. Both the behavior used by predators to capture their prey and the tactics employed by the prey to escape have therefore come under strong selective pressure in the course of evolution. Conscious thinking may be important for either predator or prey, yet most scientific descriptions of predation are couched in essentially behavioristic terms. The ethologists who have studied such behavior lean over backward to avoid any hint that the animals may have feelings or thoughts. Popular books on natural history give the impression that when prey are attempting to escape they employ only the simplest tactics, such as crouching motionless, hiding, or fleeing at top speed. Yet the evasive measures taken by potential prey, along with the tactics of predators, provide a goldmine of information about versatile behavior under natural conditions.

Laboratory studies of animal behavior almost never involve predator-prey interactions, mainly because scientists prefer not to use their captive animals as prey for carnivorous predators.

The need to standardize conditions and to repeat observations and experiments many times has also discouraged laboratory studies of predation. Predator-prey interactions are complex and variable, and they require more space than a laboratory cage. Experimental arrangements do not readily provide opportunities for the prey to escape and still allow the predator to succeed reasonably often. A further limitation is that laboratory animals have been preselected for docility.

A potential prey animal has many ways to reduce the hazard of being captured and killed. For those that live in a habitat providing protective cover, the safest behavior is generally to remain concealed. This, however, conflicts with an even more urgent necessity to gather food, so many animals are continually making compromises between locating food and avoiding capture.

Lance A. Olsen of the University of Montana pointed out to me in correspondence that an animal's self-concealment implies conscious thinking and even self-awareness. The latter is often held to be a uniquely human attribute, or perhaps one shared with the great apes (Popper, 1972; Popper and Eccles, 1977; and Bunge, 1980). Olsen has been especially impressed by the tactics of grizzly bears in seeking out positions from which they can watch hunters or other human intruders without allowing themselves to be seen (Haynes and Haynes, 1966; Mills, 1919; Wright and Kenfoot, 1909). It also has been reported that these bears make efforts to avoid leaving tracks, indicating that they realize that their tracks may be followed by human hunters. Concealment by simply moving behind something opaque would require only simple behavior patterns. But in the natural world many animals conceal themselves in such a way that they can observe the area in which some hazard has appeared or is likely to appear. A simplistic interpretation might be that the animal moves behind as much vegetation as possible while still being able to see out. But they sometimes seem to do this without exposing any part of the body to view, suggesting some such thought as "I must get *my* whole body behind something."

Gallup (1977) has conducted ingenious experiments in which chimpanzees are first given an opportunity to familiarize themselves with mirrors and then, when they are under deep anesthesia, their foreheads or ear lobes are marked with a conspicuous spot of rouge or similar material. Chimpanzees lacking experience with mirrors pay no attention to such marks, but those that are accustomed to looking at themselves in mirrors reach directly for the new spot. This seems clear evidence that they recognize the mirror image as representing their own bodies. Efforts to induce monkeys and even gibbons to use mirrors in this way have so far failed consistently, and Gallup concludes that only the great apes share with us the capacity for self-awareness. Similar experiments with gorillas by Suarez and Gallup (1981) failed to show any signs of self-awareness according to Gallup's criteria. Since it is difficult to believe that gorillas are significantly less proficient mentally than chimpanzees and orangutans, these differences between species may reflect differences in their willingness or ability to "play the game," rather than the presence or absence of self-awareness.

Because of these uncertainties, other evidence of self-awareness in nonhuman animals is important. Self-concealment may not be wholly convincing, for some will doubtless argue either that it results from the conflicting motivations to hide and to see what is happening or, if it is learned, that it has been reinforced by environmental contingencies without any conscious thinking on the animal's part. Nevertheless, Olsen's suggestions are significant. Like so many other cases of animal behavior that suggest complex thinking, however, concealment has been studied far too little to provide firm data on which to base any sort of cognitive interpretation. Hediger (1947, 1980) has discussed the indications that stags are aware of the dimensions of their growing antlers and know how large an opening they can pass through. He also suggests that animals learn that their shadows are related to their own bodies, which is a sort of self-awareness. Of course, reductionists can explain such behavior as the stag's learned response to bumping its antlers on the sides of small openings or the animal's habituation to the dark area on the ground that accompanies it on sunny days.

Many predators travel great distances watching and listening for potential prey. The actual attack and capture may happen very suddenly, and a vulnerable prey animal, such as a mouse seized by a hawk, may be quite unable to escape. While hawks, which are diurnal, see their prey, nocturnal owls usually listen for the sounds of small animals moving about and capture them by extremely precise auditory localization. Their ears and the auditory portions of their brain are highly specialized for this type of hunting. Laboratory investigations have shown that barn owls can fly at a sound source that they have learned to associate with food with great accuracy (Konishi, 1973; Knudsen and Konishi, 1978). Recent experiments by Rice (1982) have revealed that marsh hawks, or marsh harriers, also locate mice with comparable accuracy by hearing them in vegetation where they are difficult to see.

In most predator-prey interactions, however, the potential victim has some chance of escape, perhaps by outrunning the predator, moving into protective vegetation, taking shelter in a burrow or taking some defensive action such as spraying an irritating substance at the attacker.

The War between Bats and Moths

One predator-prey relationship has been studied thoroughly enough to give us substantial insights into the complex interactions that can take place even with relatively simple animals. Thirty years ago I discovered that insectivorous bats use their ultrasonic sonar not only to avoid stationary obstacles but in the much more demanding task of locating and capturing flying insects, which, I believed then, were acting as merely a passive swarm of edible droplets (Griffin, 1958). To be sure, the insects were moving, often erratically and in patterns that would seem difficult for a bat to predict. But in the early 1950s the notion that these small insects could take any sort of evasive action to avoid capture by bats seemed too far-fetched for serious consideration. But thoughtful students of insect behavior had long been aware that certain moths of the family Noctuidae react strongly, and in a puzzling fashion, to high-frequency sounds, such as the sound produced by the jangling of a bunch of keys. Key jangling often causes some species of moths to change from slow and steady flight to more rapid and erratic movements, including power dives toward the ground or vegetation.

Asher Treat of the City University of New York and Kenneth Roeder of Tufts University studied this matter carefully and demonstrated conclusively that some of the noctuid moths have specialized auditory receptors that are primarily sensitive to sounds in the ultrasonic frequency range. Similar auditory organs have since been discovered in a few other groups of nocturnal insects, located on the side of the thorax or the wings. All such organs respond to sound of relatively low intensities, and they are most sensitive to the frequencies used by insectivorous bats (Roeder, 1967; Sales and Pye, 1974; Fenton and Fullard, 1979; Miller and Olesen, 1979).

This ability to hear ultrasonic sounds enables many moths to escape from approaching bats, which emit short pulses of high-frequency sound several times per second when they are flying. As they search for insects in open air they emit sounds loud enough that the echoes from insects are audible to the hungry bat. Moths that take evasive action by flying erratically or diving into cover are far less likely to be captured, so their ultrasonic "ears" and the evasive behavior they make possible are extremely advantageous. These tactics are so effective that one may wonder why most nocturnal insects lack this auditory sensitivity and seem incapable of avoiding predatory bats. This apparently imperfect adaptation is one of the puzzles of evolutionary biology, so difficult to explain (even speculatively) that it is scarcely ever mentioned.

The ingenious and elegant investigations of Roeder and Treat included a careful physiological analysis of the auditory system used by certain species of noctuid moths. One of the two nerve cells or neurons leading from each auditory receptor to the central nervous system responds to faint sounds from relatively distant bats. Activation of this neuron causes the moth to turn away from the source of low-intensity sound. The second neuron responds only when a bat is closer and its sonar signals reach the moth at considerably higher intensities. In the species where this has been thoroughly studied, activation of the less sensitive neuron usually causes the moth to initiate a power dive toward the ground or vegetation.

Physiologists have found this a very satisfying body of data. It tickles our reductionist fancies to find that a sensory system comprising only two nerve cells on each side of the animal suffices to produce such effective antipredator behavior. Hence the whole picture of how moths avoid bat predation has become a sort of classic example of comparative physiology. But no one has inquired what the bats and moths might be thinking during these interactions. Since the moth's evasive behavior is initiated by only four sensory nerve cells, we may think that it is unlikely to require anything remotely like conscious thinking. But the moth's central nervous system is far more than this limited input channel; coordination of flight requires hundreds or thousands of neurons. Although the neural mechanisms controlling complex behavior in moths have been studied only to a limited extent (Surlykke and Miller, 1982), experiments with locusts have revealed that interactive networks of neurons and synapses regulate behavior in much the same way as in mammalian brains (Hoyle, 1977). As pointed out in Chapter 2, even the simple act of moving away becomes complex when we describe it in neurophysiological terms. If moths can feel anything, this is a time for terror and conscious striving to escape.

Bats must skillfully adjust their flight maneuvers to capture their prey, and they often succeed despite the moths' evasive tactics. Within half a second the bat may reach out with a wingtip or the tail membrane to enfold the insect and transfer it to the mouth. When insect prey is abundant, the bat repeats these elaborate maneuvers every few seconds and can fill its stomach with a quarter of its body weight in less than half an hour. It does this by making a series of individual pursuits, not by flying at random with an open mouth. What does the hunting bat think about? Probably not about every wingbeat or sonar echo; my guess is that the bat thinks about the taste of the insect prey rather than the maneuvers required to catch them. If inedible insect-sized objects are thrown up into the air, bats will pursue, catch, then hastily drop them; but when this trick is repeated many times, they eventually learn to distinguish the edible insects from similar objects. Under natural conditions they distinguish among various kinds of insects that are locally available, as demonstrated by Buchler (1976) and Goldman and Henson (1977).

Hunting and Escaping on the African Plains

Because they are fairly easy to observe, we know quite a bit about the behavior of diurnal mammals living on the open plains of East Africa. Their visibility has allowed diligent ethologists to learn how predators hunt and how their prey often escape capture. The work of three ethologists is particularly relevant to the question under discussion in this book. Fritz R. Walther (1969) studied in detail the most abundant of the many species of antelopes, the Thompson's gazelle, which are commonly known as "tommies." Hans Kruuk (1972) spent several years studying the behavior of the spotted hyena, one of the most abundant and effective predators of the East African plains. Finally, George B. Schaller (1972) concentrated on the lions of the Serengeti Park in Tanzania. From these three investigations, supplemented by the work of several other scientists, it is possible to describe the principal hunting tactics of two predators, the hyena and the lion, and the responses of their prey, tommies and other antelopes.

Popular accounts of the animals of the East African plains

suggest an erroneous picture of all-powerful predators surrounded by an endless sea of potential food; whenever they are hungry they need only seize the nearest morsel. A comparable fallacy is that grazing animals live in a state of constant terror and survive only by good fortune when no lion or hyena happens to come their way. But these prey animals do not always flee when they first detect a lion or hyena, and they seem to understand a great deal about the dangers presented by these predators. Of course, many are captured, but not so many as to prevent successful reproduction and perpetuation of their populations.

The Victim's View

It is helpful to begin by concentrating on the tommies' behavior. In many ways the most thoroughly studied of the East African grazing animals, Thompson's gazelles are about the size of a large dog and typically weigh 10–15 kg. They congregate in herds that vary in composition according to the circumstances. During migrations males and females may be about equally abundant in mixed herds, but when they remain in one area for a time most herds consist of females and so-called bachelors. The adult bucks occupy and defend individual territories 100 to 200 meters in diameter, with relatively little space between territories. Herds of females move through these territories and are tolerated even when courtship is not occurring. The bachelor males, on the other hand, are chased away and are found predominantly around the edges of these territories.

As Walther emphasizes, tommies escape from lions or hyenas principally by running rapidly away. Such flight is critically important, but it is surprising how much of a tommy's life is spent in apparent peace and quiet, even when predators are close enough to be clearly visible. Ordinarily the tommies seem less disturbed by predators than by a heavy rainstorm. The most common, indeed an almost continuous, type of predator avoidance behavior is simply looking around, yet a grazing tommy may not look around for as long as fifteen minutes. In a large herd the looking around is far from synchronous, so most of the time at least one tommy is likely to be alert to any approaching predator. On seeing something conspicuous or suspicious, the tommy's posture becomes alert, with the head held high, ears somewhat forward, and muscles tensed. Sometimes it stamps a foreleg, but not always.

When one tommy assumes an alert posture, the others quickly notice. Sometimes the alarmed gazelle makes a soft snort, which Walther describes as sounding like "quiff," a sound that seems not to be used in other situations and apparently conveys a message like "There's something scary in the direction I'm looking." The rest of the herd immediately turn their heads to look in the same direction as the alarmed one. When they first detect a predator or anything unusual or conspicuous, the tommies often do not flee but move closer. This behavior sometimes seems to suggest that they are fascinated. Walther observed that a herd of tommies would occasionally recognize a predator at 500 to 800 meters but would then approach within 100 to 200 meters, with the individuals staying close to one another. If the predator moved, the herd would follow it. The gazelles are apparently aware of the predator and ready to run off if it shows signs of attacking. The predators also seem to understand this situation and rarely rush at a group of alert tommies.

Predator monitoring is especially striking in the case of territorial males. During the daytime, if a predator enters the territory of a male tommy, the females generally move out of the territory. The male usually remains and keeps the predator under close watch. As the predator moves, the buck usually follows it until he reaches his boundary, at which point his territorial neighbor ordinarily notices his alertness and starts to monitor the lion or hyena. The effectiveness of predator monitoring by territorial males was demonstrated by Walther's intensive studies of their social behavior. He observed more than fifty until he knew them individually and watched them almost continuously for months at a time. Only one was taken by a predator during two years of observation.

Walther observed an especially intriguing relationship among predators whose home area was close to that of territorial tommies. Hyenas ordinarily rested from midday until late afternoon in their dens, which are often within the territories of the male tommies. In the early evening when the hyenas left their dens, they were surrounded by tommies, but they never tried to hunt one. Instead they passed between them and hunted in other areas. It might be that the local resident gazelles knew all about the hyenas and were unlikely to be caught off guard. Or perhaps the hyenas found the resident tommies familiar enough to refrain from hunting such neighbors, or the hyenas may have been reserving them for times when no other prey was available.

When a lion or hyena makes a serious hunting attack, the tommies run at a steady gallop, attaining speeds of 45 to 60 km per hour, faster than any of their predators can run except for a very short distance in an initial charge. But the tommies use other gaits, including one called stotting in which they jump a meter or so into the air, holding the legs relatively stiff. Stotting occurs most often at the beginning of a chase, if the pursuer is not too close, and at the end, after the pursuer has given up. In a hotly pursued herd most or all of the gazelles are bouncing about in an irregular fashion, which seems to confuse many predators and prevent them from concentrating on a single potential victim so that the entire group often escapes. But galloping is much faster than stotting, and when closely pressed the tommies always try to escape at a gallop. Tommies react differently to different predators, stotting frequently when chased by hyenas or wild dogs but only rarely for lions, cheetahs, and leopards, which can maintain a higher speed in an initial rush.

Walther describes at some length a number of behavior patterns exhibited by tommies just before their actual capture by predators. In these desparate circumstances they often change direction rapidly, doubling back like a hare, but in all cases where Walther observed such behavior the predator captured the gazelle in the end. When approached by a jackal, which is about the size of a fox, a mother and fawn would at first run off together, although sometimes the mother would attempt to defend the fawn by charging at the jackal and striking it with her horns. Occasionally a second female, quite possibly an older sister of the threatened fawn, joins in this kind of attack. Mothers with young fawns may attack quite harmless animals and birds that do not offer any real danger. But when a fawn is being chased by a larger and more dangerous predator, such as a leopard or cheetah, the mother moves about excitedly at a considerable distance without intervening.

Hans Kruuk describes a sort of distraction behavior on the part of a mother tommy when a hyena is chasing her fawn. The mother may run between the two, crossing in front of the hyena, staying very close beside it and yet just out of reach. Usually only one female does this at a time, presumably the fawn's mother. But Kruuk occasionally saw as many as four female gazelles acting in this way at the same time to help one fawn. Again, all of these efforts seem to be ineffective. The distracting adults do not seem to have much effect on the hunting success of the hyena; six out of twelve fawns were captured despite the efforts of the adults, whereas in nineteen cases where the females did not try to distract, the hyena captured only five.

Behavioral ecologists speculate that such apparently hopeless behavior cannot have been favored by natural selection, and they seem puzzled that it occurs at all. However, if we entertain the possibility that a Thompson's gazelle is fully aware that it is in great danger of being caught by a pursuing predator, we are not surprised that it makes every attempt to escape or at least postpone an inevitable fate. Likewise, mothers may want to distract a predator and save their young even when, on evolutionary grounds alone, this would appear to be wasted effort at best and maladaptive at worst. Such close approach to a predator certainly entails some risk, and occasionally a mother behaving in this way is captured.

George Schaller also describes potential prey monitoring the behavior of visible predators and not appearing frightened unless the predator rushes at them at high speed. If a lion is walking along steadily, tommies, zebras, wildebeest, or other potential prey ordinarily face the danger in an erect posture but do not run away. Wildebeest usually keep up an incessant grunting, but when a lion approaches, they stop, creating a zone of silence around the predator, which undoubtedly helps warn the other wildebeest. Prey animals often approach a predator, and a group may even line up to watch it pass. If the lion stops and turns in their direction, the grazing animals may flee for a short distance, then turn and stand watching again. Roughly 30 meters from a lion seems to be considered a safe distance in open country. On the other hand, when potential prey animals move into thick vegetation, they become much more cautious. In other words, they behave sensibly.

The Predator's View

Predators also monitor the behavior of potential prey. Hyenas are especially alert for slight differences in an individual's locomotion or other behavior that may indicate that it is vulnerable and can be captured more easily. Hyenas will often attack an animal that has been temporarily anesthetized by scientists for study or marking, even when the animal has apparently recovered and is moving normally. The investigator who wishes his study animal to survive must intervene; Kruuk would stand by in his

Land Rover to chase away attacking hyenas. Hyenas selectively attack any weak, sick, very young, or very old members of a herd, but when all appear perfectly healthy and vigorous, they still manage to capture some.

Kruuk remarks on a number of ineffective and apparently useless forms of behavior exhibited by wildebeest. Although wildebeest are roughly the size of a cow, they seem singularly inept at defending themselves when surrounded and attacked by a pack of hyenas. They make only faint attempts at butting their attackers, and as Kruuk describes the situation: "Generally speaking, the quarry just stands uttering loud moaning calls and is torn apart by the hyenas. It appears to be in a state of shock" (p. 158). Sometimes a wildebeest runs into a lake or stream, but the hyenas almost invariably kill it. Here again the wildebeest may perhaps hope, albeit vainly, that it can escape by running into the water.

Kruuk noticed that when a hyena is hunting a wildebeest calf, the mother often will attack the hyena, apparently to protect the calf. Wildebeest cows without young ordinarily refrain from any such attack, but Kruuk observed one striking exception. A cow in the process of giving birth—the calf's front feet had already emerged—went to some lengths to attack a hyena walking past that seemed not to be paying any attention. Shortly afterward the cow seemed to be avoiding the other hyenas in the vicinity and, selecting a quiet place surrounded by other wildebeest, lay down to complete the birth. Presumably a cow in advanced labor is not in prime physical condition to engage in combat, and the calf is not yet available to be attacked. Might the cow realize that hyenas are a danger to her soon-to-be-born youngster? Reductionists might argue that hormonal levels, perhaps connected with impending lactation, caused the female wildebeest to depart from her normal behavior and attack the hyena. But we should recognize how little we know and keep our minds open to the possibility that wildebeest mothers have simple thoughts like protecting their babies and attacking predators they have seen kill other young wildebeest.

From these extensive observations there emerges a general picture of the relationships between predators and prey on the African plains. Predators watch the behavior of their potential prey, and the herbivorous animals keep an eye on visible predators and watch for those that may appear. But only on relatively rare occasions does a predator become serious about attacking and only then, ordinarily, do the prey animals become seriously alarmed and take strenuous actions to escape. Both predators and prey spend most of the time when they are visible to each other in monitoring rather than in attack or escape. Such monitoring involves making subtle distinctions; predators notice when prey animals behave abnormally, since this often indicates weakness and vulnerability, and the prey are very alert to subtle behavior changes that indicate whether or not the predators are a serious threat. In this kind of situation, in which many factors must be weighed and evaluated, conscious thinking would be effective, and it may well occur more often than we usually suppose.

Cooperative Hunting

An especially significant situation is when two or more lions, hyenas, or other predators can improve their chances of success by combining their efforts. Sometimes the advantage of group hunting is in being able to overcome larger prey, as when a group of wild dogs or hyenas kills an animal such as a wildebeest or zebra that is large enough to fight off a single attacker. Like wolves in colder climates, hyenas and wild dogs seldom waste their efforts in individual attacks on much larger animals.

To what extent does group hunting involve cooperation, and does this involve conscious planning? Most of the time the behavior of predators hunting in groups does not show obvious signs of coordination, although sportsmen and casual observers of animals have often taken it for granted that intentional cooperation occurs. But cautious ethologists have concluded that each carnivore operates independently, and that benefiting from the actions of a companion is more or less accidental.

For example, Hans Kruuk (1972) summarized extensive observations of the tactics used by hyenas in hunting wildebeest calves. Out of 108 attempts by one or more hyenas to capture calves, only about one-third were successful. Single hyenas rarely succeed, and in many cases the mother's active attacks save the calf. When two hyenas attack a calf, however, the mother is often unable to drive off both of them; while she is attacking one, the other seizes the calf. Many observations of hyenas and wild dogs attacking tommies and other prey make it clear that group hunting is more effective than the efforts of single animals. But does this involve cooperation beyond simply taking advantage of the greater vulnerability of the prey?

Some of the most helpful information about cooperative hunting has resulted from studies of lions, especially those by George Schaller (1972). He occasionally observed that a group of four or five lionesses would spread out as they approached one or more gazelles; those in the center would move more slowly than those at the edges, creating a U-shaped pattern. By the time any lion was close to the prey, the latter were surrounded by lions on three sides. When all the lions rushed at once, the chance was clearly greater that at least one would succeed than if they had all approached in a line or in a dense pack. Kruuk and Schaller do not conclude from such behavior that the lions are cooperating in any intentional fashion. The index of Schaller's book *The Serengeti Lion, A Study of Predator-Prey Relations* does not contain the word cooperation. Although Schaller devotes three pages to communal hunting, he concludes that each lion behaves more or less independently but is quite ready to take advantage of another lion's hunting efforts. When a tommy is fleeing from one lion and comes within rushing range of another, the second is far more likely to succeed than would otherwise be the case. Schaller also describes how lions in a group approaching prey look at their neighbors and seem to maintain an effective position, such as the U-shaped pattern. He feels that although such cooperation is real, it is limited to maintaining appropriate positions in relation to each other. But even such spacing implies that the animal understands the advantages it affords.

Schaller, Kruuk, and other students of predator-prey relations thus go to considerable pains to deemphasize the possibility that lions and other predators consciously collaborate. They are reflecting a scientific caution that arose in reaction to uncritical assumptions that predators must cooperate as human hunters often do. Nevertheless this caution may have been carried too far, and suggestive evidence of cooperation may have been given less weight than it deserves.

During a short visit with the ethologists Robert Seyfarth and Dorothy Cheney Seyfarth at Amboseli National Park in Kenya, I had the good fortune of seeing what looked very much like a cooperative lion hunt. During the last hour of this two-day visit, the Seyfarths drove me along a dirt road at the edge of an open plain close to the forest boundary. At a place where the edge of the woodland receded for a few hundred meters, the road ran between a semicircular area of grassland and open plain. A large herd of wildebeest had split into two groups; fifty to sixty were grazing on the grass on the woodland side, while the remaining hundred or so were feeding on the open plain about 150–200 meters from the road.

We had scarcely paused to watch the wildebeest when five lionesses approached with a businesslike gait along the plain roughly parallel to the road. Both groups of wildebeest clearly saw them, stopped feeding and watched intently. From our position on the road we could not see the lionesses all the time, since the ground was irregular, but when they were about 200 meters from the two groups of wildebeest, two lionesses climbed slowly to the tops of two adjacent mounds. They then sat down and remained stationary but conspicuous. After a few moments we could see a third lioness slinking, her belly pressed close to the ground, through a ditch that paralleled the road. Although we could see her only occasionally, she was clearly moving toward a position roughly midway between the two groups of wildebeest. She soon crawled out of our view and for several minutes nothing seemed to be happening at all.

Then suddenly a fourth lioness, almost certainly a member of this group, rushed out of the forest behind the wildebeest on that side of the road. Apparently she had moved out of our view, and presumably that of the wildebeest, to a position ideally situated to chase them across the roadway toward the other group on the open plain. They thundered off in precisely that direction, crossing the road and the ditch just about where we had seen the lioness sneaking furtively a few minutes earlier. As the herd bounded over the ditch, that lioness leaped up and seized one of the fifty or so that were galloping all around her. As the dust settled she began killing the animal by covering its mouth with hers as it lay on its back, legs kicking gently in the still dusty air. The other four lionesses walked in a very leisurely fashion toward the downed wildebeest but arrived only after its kicking had ceased. The group ate their prey in a leisurely way, and after a few moments a jackal joined the scene. The wildebeest and a few zebras returned from the open plain and stood in a line watching the lionesses and their victim from a distance of about a hundred meters.

This single observation cannot be taken as conclusive proof of intentional cooperation, but it was certainly very suggestive. Why should two lionesses climb to conspicuous positions where the wildebeest could easily see that they presented no serious danger? Why should a third sneak along the ditch to a position about midway between the two groups? Was it a pure coincidence that a fourth lioness just happened to rush out from an optimal point at the forest edge to chase the wildebeest over the ditch where one of her companions was waiting? Students of African animals have confirmed that we were remarkably fortunate to see so much of this sequence of group hunting behavior. Other ethologists have observed elements of the pattern I have described, yet they seldom write about such observations, probably for fear of seeming unscientific.

Considering the obvious advantages to cooperative hunting, it seems reasonable to conclude that lions are capable of planning their hunting tactics. As with any human behavior, even the most complex and thoughtful, reductionist explanations are possible. Perhaps the genes of lions, molded by natural selection over thousands of years, contain instructions that guide the details of this and several other patterns of coordinated behavior involved in cooperative hunting. Undoubtedly, adult lionesses have learned a great deal both from watching their mothers, sisters, or other companions hunt, and from their own efforts. What the inclusive behaviorists have ignored is the possibility that when cooperation is effective, predators may consciously think about what they are trying to accomplish. It seems foolish to assume that no conscious thinking ever occurs when groups of predators coordinate their hunting efforts. Clearly we need more evidence, but we can consider that in such situations simple conscious thinking is quite likely.

Predator Distraction

Certain species of birds engage in some especially important types of antipredator behavior when their eggs or young are threatened by large, possibly dangerous intruders. They may simply fly away, but even in that case, to draw the predator away from the nest they must make quiet, inconspicuous movements while still close to the nest. Before considering apparently intentional distraction, I should emphasize that when birds are incubating eggs or caring for nestlings they often attack approaching predators. They make loud and often raucous calls, fly at the intruder, regurgitate their stomach contents or empty the cloaca at the intruder, and sometimes strike it with bill, legs, or wings. Such attacks are often surprisingly effective, even against larger birds, including hawks, which are quite capable of capturing their attackers. Blackbirds, kingbirds, and other relatively small birds often attack crows, which they seem to recognize as dangerous predators on their eggs or young. The smaller birds usually fly above the crow and dive at it, and ordinarily they succeed in driving it off. Even fast-flying hawks that capture small birds on the wing are sometimes attacked, and these attacks resemble mobbing in many ways. A group of small birds will mob a hawk or owl, flying about noisily and making a great nuisance of themselves to the larger predator. Mobbing seems to occur at any season when numerous small birds detect a hawk or owl. They dart at it, almost pecking it on occasion, and usually cause it to move away. Occasionally, however, a hawk does turn on one of its tormenters and kill it, so mobbing is not without risks. Unlike mobbing, predator distraction is almost always the work of parent birds with vulnerable eggs or young, although other adults of the same species may join them.

It is not surprising that parent birds should become aggressive toward predators that are a danger to their eggs or young. It seems reasonable to infer some primitive sort of rage, fear, or a combination when a relatively small bird attacks a larger bird or mammal that is approaching its eggs or young. Can we also infer that the bird anticipates various possibilities for the near future? Does such a parent think, "If I don't drive that creature away, it may kill my youngsters" or "Stay away from my nest, you——?" Rage or the motivation to attack another, usually larger, creature is not a very complex feeling. Certainly birds show signs of such angry aggression against larger animals far more often when they are caring for eggs or young than at other times. They will vigorously attack or scream at predators from which they would ordinarily flee or watch from a respectful distance. Whatever feelings the bird experiences, its behavior depends on whether or not it is raising young.

In addition to outright attack, several species of ground-nesting birds, especially the smaller plovers and sandpipers, behave in a more complex way that strongly suggests they are intentionally leading a predator away from their vulnerable young. Most of these birds nest in the high Arctic, so we are not familiar with their behavior during the breeding season. But in temperate latitudes the killdeer, which nest on open fields, and piping plovers, which lay their eggs on sandy beaches, very commonly display this predator-distraction behavior.

Ornithologists and ethologists have repeatedly observed the behavior of nesting plover when a large intruder, such as a person, approaches a nest where a killdeer or piping plover is incubating its eggs. At a considerable distance, long before a human observer or other mammal can see the cryptically colored bird or its eggs, the plover may stand up and walk slowly to a point a few meters from the nest. Only then does it begin the plaintive calling that gives the piping plover its name. The bird may then walk rapidly or fly in almost any direction except toward the nest. If a person approaches these birds while feeding or when they have no eggs or young, they fly away from the intruder to a safe distance, perhaps resuming their search for food. When they have no vulnerable eggs or young, the plovers almost never approach an intruder or act in a way that makes them more conspicuous.

Most of the scientific observations of predator distraction display have involved human intruders, usually the observer himself. But on a few occasions the same sort of behavior has been seen toward wild carnivores that might well eat the eggs or young. Several ornithologists have used trained dogs whose approach could be generally controlled, and on a few fortunate occasions a careful observer has watched the interaction between a nesting bird and a fox, otter, skunk, or weasel. The behavior seems to be similar in all these cases, and it is usually effective.

The bird flutters slowly but conspicuously away from the nest, staying relatively close to the intruder. It almost always makes loud piping or peeping sounds similar to those a bird makes when disturbed or mildly irritated. It may display conspicuous feather patterns that are not usually visible. If the bird is walking or running, its gait is different from normal locomotion and more noticeable. Sometimes the bird uses a so-called crouch run, with the body held closer to the ground and head lower than normal; its running is almost like that of a small rodent. Some species even make somewhat rodentlike squeaks when running this way.

It is common for the bird to hold its tail or wing in an abnormal position as it moves. Often the tail almost drags on the ground, and the wings are slightly extended, sometimes one more than the other, strongly suggesting some weakness or injury. After running a few meters, the bird may flop about on the ground, extending one or both wings, as if injured. This is often called the "broken-wing display," and it requires considerable effort for an observer to believe that the bird is really quite healthy. As mentioned earlier, predators are extremely sensitive to minor differences in the gait and demeanor of potential prey and are much more likely to attack animals that are behaving abnormally.

Throughout most of this predator-distraction behavior, the bird watches the intruder. Typically it does not move in a straight line and stops from time to time. If the intruder approaches, the bird moves farther ahead. If not, the bird usually flies back closer to the intruder and repeats the behavior. The bird will allow the intruder to approach quite close, sometimes within two meters, but it always moves just fast enough and far enough to avoid capture. Typically the bird continues the injury simulation while leading the intruder some distance away from the nest or young. Finally, however, it flies away rapidly, usually in the same direction, then circling back to the general vicinity, though seldom to the exact spot, where the eggs or young are located. One Wilson's plover, a close relative of the killdeer and piping plover, led me more than 300 meters along a sandy beach before flying off.

Ethologists have customarily explained predator-distraction behavior as an instinctive action involving conflict between two opposing motivations, fear of the predator and motivation to attack. Since these call for moving in opposite directions relative to the predator, a sort of uncoordinated convulsion is said to result, reducing the bird to random and meaningless activity. It has even been claimed that such birds are virtually paralyzed.

Anyone who has carefully watched the predator-distraction display of a small plover will find the concept of a paralytic convulsion quite difficult to reconcile with what he sees. The bird is clearly controlling its behavior and modifying it in detail according to what the intruder does. It looks frequently at the intruder, continues in one direction if the intruder follows, but flies in a well-coordinated fashion back to the intruder's vicinity if he does not. Furthermore, if the intruder comes too close, the bird always recovers its coordinated locomotion. There are many well-orchestrated complexities to the behavior, and its adjustment to circumstances strongly suggests intentional reaction to the situation rather than crippling confusion.

Predator distraction also varies according to the nature of the intruder. Flying birds of prey almost never elicit this sort of behavior. The parents attack them, flee from them, or simply stay out of their way. The behavior I have been describing ordinarily ensues when a mammal approaches, including a human intruder. It may not be highly selective, and sometimes a nondangerous creature elicits some elements of predator distraction. But it does seem to be more intense at the approach of a truly dangerous intruder. It has been reported that when cattle or horses approach, the killdeer will stand close to its nest, spreading its wings and making itself as conspicuous as possible, or it may charge. This behavior pattern commonly causes the hoofed animal to turn to one side and avoid stepping on the nest. It is not known whether these birds had experience with different sorts of mammals before these observations were made. It may be that cattle or sheep elicited typical predator distraction behavior at first, but after many experiences in which the animal neither responded to being led away nor ate the eggs or young, the parent bird's behavior changed to a conspicuous display or attack close to the nest (Skutch, 1976 (p. 408); Armstrong, 1949).

Habituation is very important in animal behavior; nesting birds often engage in less and less predator distraction behavior when people or other intruders come near the nest repeatedly but do no harm. What is needed are long-term and careful observations in which the behavior of parent birds toward a variety of intruders is studied thoroughly enough that one can test the validity of this explanation. But even if the difference in behavior directed at dogs and people, on the one hand, and cows and sheep, on the other, has resulted from learning and habituation, this would show that the behavior is varied appropriately according to circumstances.

Several other aspects of predator distraction behavior are even more suggestive. As mentioned above, an incubating plover often notices an approaching intruder at a considerable distance and moves several meters away from its nest before using an abnormal gait or simulating injury. Some plovers then make a small depression in the sand and squat as if they were incubating eggs. The predator is drawn toward the apparent nest and away from the real one. False incubation ordinarily occurs at some distance from the nest when a predator is approaching, but it has also been observed under other circumstances, so it is not rigidly linked to distracting intruders. Sordahl (1980, 1981) has recently described how avocets and stilts respond to an approaching predator with false incubation; they also employ the broken-wing display in other circumstances. The great subtlety and variability of predator-distraction display suggest that the birds must have some understanding of what they are doing and the likely results— namely, preventing the destruction of their eggs and young.

Predator distraction is rare or feeble before eggs are laid, increases during incubation, and is usually most intense and prolonged after the eggs have hatched and the young are vulnerable because they cannot yet fly. In many species the newly hatched young move about, joining the parents in feeding or being fed by the parents. Under these circumstances the parents often give faint calls when an intruder approaches, which induce the young to hide. Then the parent may carry out predator distraction. When the young can fly, this behavior becomes sporadic or stops altogether.

Most of my descriptions have been couched in quite objective terms, and I have tried to avoid any interpretation of the parent birds' possible feelings or thoughts. Fear and anger seem highly likely, specifically, fear that the intruder will hurt or kill the eggs or young. But do they go further and think something like "I will lead that horrible creature away from my young"? Their behavior is quite consistent with such an interpretation, but skeptical scientists have long objected that this is not sufficient evidence to infer even simple thoughts. But consider how these birds commonly behave in other respects. When an intruder approaches, one might suppose that a fearful parent bird would move away from the intruder, but it tends to do the opposite. Perhaps this first phase of predator-distraction behavior is a sort of low-intensity attack, but in that case, one might suppose it would be more likely to fly close to the intruder. Walking slowly to one side and displaying false incubation are scarcely actions that suggest angry aggression.

The amount of effort devoted by parent birds to monitoring the intruder's behavior also suggests the intention of leading it away rather than oscillation between angry attack and fearful flight. Interrupting the crouch-run, tail-down posture or broken-wing display to peer at the still-distant intruder suggests that the bird wants to know whether the intruder is following. And if the intruder does not follow, the bird usually ceases its abnormal behavior and flies closer before starting a similar pattern in a new location. Does the bird entertain thoughts such as, "That creature isn't following me, so I will fly back and try to lead it off in some other direction"?

A final difficulty with the explanation based solely on conflict between attack and flight is that if the intruder does follow closely, the parent bird continues its abnormal behavior for some distance, then suddenly resumes its normal flight, leaving the intruder altogether. According to the conflict explanation, one must suppose that the entire conflict has suddenly been resolved, even though the threatening intruder is closing in.

Predator-distraction behavior is almost never interpreted by ornithologists or ethologists in terms of feelings, thoughts, or intentions. On the contrary, inclusive behaviorists make every effort to avoid such concepts, appealing to presumed evolutionary selection of behavior patterns that have resulted in more abundant reproduction by the birds' ancestors. But these alternative explanations are by no means mutually exclusive. If birds experience simple feelings, thoughts, and intentions, they may well occur together with conflicting motivations. Such thinking might accompany behavior patterns that are carried out because of the bird's genetic makeup, as suggested in the final section of Chapter 2. Thus, to infer that parent birds are influenced by conflicting motivations does not preclude the possibility that it is thinking about the situation. The standard interpretation may well be partly correct, but this is quite a separate matter from the question of whether the bird experiences subjective feelings or conscious thoughts.

Why has conflict between motivations been so commonly accepted as an adequate explanation and, at least by implication, one that does away with any need to suppose that the bird has the slightest idea what it is doing? Perhaps this preference tells more about the scientists than the birds. If we pull ourselves out of this negative dogmatism, we can begin to ask what birds engaged in predator-distraction behavior might be feeling and thinking. Furthermore, as scientists and thoughtful inquirers into the nature of the living world, we can consider how we could learn more about whether animals have conscious thoughts.

If we begin with the common situation where an intruder has been spotted at some distance, let us imagine that the bird already has a fear that this creature will harm its eggs or young. The bird might also think about two possibilities: either that creature will come to my young and hurt them, or it will go off in some other direction and leave them alone. If the bird considers these two possibilities, it will obviously prefer the second. As it slowly stands up and walks away, it might think, "I can make that intruder go away from my young." The display of false incubation fits into this simple expectation. "If that creature comes closer, it may see me rather than my youngsters."

Suppose now that the intruder does come closer and that the parent is still thinking about these two possibilities. It arises from its false nest and moves toward the intruder, using an abnormal gait or flight. It may well be thinking, "Perhaps that beast will follow me." If the intruder nevertheless walks closer to the nest or young, the bird may shift to the crouch run or the broken-wing display. If the intruder still approaches, the bird may think something like, "He is coming toward me and I will lead him this way, not toward my nest." Watching the intruder is clearly important, and continuing the leading-away behavior or stopping it could be accompanied by the thought "He is following me" or by a realization that the intruder is still moving toward the nest and that another attempt is required.

The thoughts I am ascribing to the bird under these conditions are quite simple ones, but it is often taken for granted that purely mechanical, reflex-like behavior would be a more parsimonious explanation than even crude subjective feelings or conscious thoughts. But to account for predator distraction by plovers, we must dream up complex and tortuous chains of mechanical reflexes. Simple thoughts could guide a great deal of appropriate behavior without nearly such complex mental gymnastics on the part of the ethologist or the animal.

If we postulate, tentatively, the presence of simple thoughts, a great deal of animal behavior can be understood as a consistent adaptive pattern. Explaining all the variations in predator-distraction behavior solely on the basis of conflict between motivations requires numerous separate and ad hoc assumptions to account for false incubation, the rodentlike gait, making oneself conspicuous to the threatening predator (but not near the nest), and leading away from nest or young. But if we recognize that the parent bird simply wants the predator to move away from its young, the behavior patterns fall into place as reasonable procedures to achieve a vitally important goal.

5 Artifacts and Templates

Many animals alter their immediate environments in ways that increase their chances of long-term survival and successful reproduction. Doing this usually requires complex and lengthy efforts, and the structures they build are entirely different from anything produced by nonbiological processes. In certain cases building behavior suggests conscious thought about the advantageous results to be attained. Although we can never be absolutely certain about what goes on in the minds of animals as they build nests or burrows, or make arrangements to store food, it is clear that in some cases thoughtful anticipation would facilitate these endeavors. I will use the term artifact to denote objects that animals work on for the benefit of themselves or their offspring or companions. In Chapter 6 I will consider those artifacts that are carried about and used as tools to achieve results not readily accomplished by the animal's appendages or mouthparts. While there is some ambiguity in this distinction between artifacts and tools, I hope it will help to avoid entanglement with semantic questions of definition that have encumbered other discussions.

Scientific studies of the structures animals build are widely scattered, and many are simply descriptive. For instance, the nests of termites and the cases of caddis-fly larvae are often described and illustrated primarily to aid in species identification. But Karl von Frisch has assembled many significant examples in a delightful book, *Animal Architecture* (Frisch, 1974). It is no accident that the same man who discovered the symbolic communication of honeybees, discussed in Chapter 9, also turned his attention to the shelters constructed by various animals. Much of this chapter is based on the material reviewed by von Frisch; I recommend his book to readers who want to find out much more about animal architecture.

Modifications of the Environment

Almost everything an animal does has some effect on its environment, but for our purposes these rather general effects are of less interest than the specific, often complex modifications of the immediate surroundings carried out by many types of animals. Then actions extend over time, and the results are important to the animal in the future. Thus, anticipation of future benefits could help the animal modify its environment in appropriate ways.

When small mammals walk repeatedly over more or less the same route, they wear down the vegetation, compact the soft ground, and form a distinct pathway, making locomotion easier and more rapid. In many places small animals make more or less covered runways by removing only the vegetation closest to the ground while leaving in place the plants directly overhead. The runways of meadow voles in grassy areas are a good example. Even in dense grass, the floor of a vole runway is smooth and level. The runway may have started as an incidental result of repeated walking over the same route, but its users soon work on it actively, nibbling away at the lower parts of some plants while leaving in place the blades of grass that lean over the runway. Such a runway is almost invisible from above and can be used as a miniature highway for the voles. Some rodents construct partly or wholly concealed runways between their nests or shelters and their feeding areas, greatly reducing their vulnerability to predators when they are traveling between the two locations. Animals do not build runways at random but only between places that are important to them.

Construction of Shelters

An amazing variety of animals construct shelters of one sort or another. The most familiar examples are the nests of birds and the burrows of mammals, but many insects construct even more elaborate shelters. Although one ordinarily thinks of fish and other aquatic animals as leading shelterless lives, simple nests are built by many fishes (Norman, 1949) and by several aquatic invertebrates (Frisch, 1974) either to protect themselves or their eggs and developing young. Even some of the simplest animals, such as single-celled protozoans, construct shelters of a sort, which obliges us to inquire whether we can make a distinction between shelters that are likely to require conscious thinking in their preparation and those that are not.

To begin at an extreme where conscious thinking seems impossible, some species of single-celled amoebas are covered by layers of hard protective material for most of their lives. In most cases the cell synthesizes the material; it is either secreted on the cell surface or arranged into elaborate internal structures partly or wholly surrounded by protoplasm. Most of the very abundant Foraminifera have such skeletons, which form complex and often beautiful shells. Pseudopodia or temporary projections of the protoplasm emerge through openings in the shell for food gathering and other exchanges with the environment. In a few species of amoeba-like protozoans, the protective coating is formed from minute particles of sand or mud. One of the common protozoans of this sort, *Difflugia,* looks like a miniature croquette (Hyman, 1940).

How do *Difflugia* and similar Protozoa acquire their outer crust? Ordinarily the particles are engulfed by the protoplasm, much as particles of food are, then move through the cell to the outer surface where they remain. In early studies of similar protozoans, German investigators found that under certain chemical conditions an oil drop immersed in water would acquire a crust of particles that are attracted to the interface of the oil droplet and the water. The particles tend to repel each other and thus are distributed fairly uniformly over the surface of the droplet. Under suitable conditions *Difflugia*-like "croquettes" could be produced by such purely physical processes, but it is not certain that this is the entire explanation of the naturally occurring particles (Netzel, 1977). The cellular processes do seem to be comparable to the orderly patterns of growth and cell differentiation rather than to behavioral choices. It does not seem reasonable to suppose that an amoeba thinks and plans its acquisition of protective particles, because protozoans do not possess anything remotely like a central nervous system in which information can be received, sorted, integrated, stored, and utilized in thinking or feeling.

Caddis Fly Cases

The behavior of another group of aquatic animals that construct protective coverings is more suggestive of intentional action. These are the larvae of caddis flies, which are quite abundant in fresh water streams and ponds. Superficially the larvae are rather like the familiar caterpillars that metamorphose into moths and butterflies. Like caterpillars, they have versatile mouthparts for cutting up particles of vegetation or capturing smaller aquatic animals (Wiggins, 1977). But they develop from eggs laid in the water by the winged female caddis flies, which emerge and mate after a long aquatic larval period.

The larvae of almost all species of caddis flies cover their bodies with particles of sand, bits of leaves, or other available particles, cemented together by silk secreted from glands on the head. In the early larval stages, the animal uses small, homogeneous particles to form a roughly cylindrical case, often with a single particle closing the front of the case. These cases almost certainly protect the otherwise vulnerable larvae from predation by small fishes or predatory larvae of insects such as dragonflies.

The caddis fly case is not totally impervious; a hole at the posterior end allows feces to pass out, and water circulates freely through the case, so the larva's gills can extract oxygen. When the larva moves about, it pushes its head and the thoracic segments bearing the six legs out through the front opening; small hooklike projections on the abdominal segments hold the case close to the body.

There are several dozen genera of caddis flies throughout the world, and their larval cases are enormously varied and sufficiently characteristic that they are often identified more easily by their cases than by the structure of the larva itself. The larvae are somewhat selective about the materials used in case construction, within the limits of what is available. Many cases consist of grains of sand or mud or other inert material, but some species cut pieces from the leaves of aquatic plants and a few use the discarded shells of tiny aquatic snails. Some species build structures that serve to capture prey from the flowing water; the case may simply be enlarged at the upstream end, or the larvae may build nets of parallel strands of silk that strain out minute aquatic plants and animals. In the few species that have been studied carefully in the laboratory, the larva can be seen to pick up a variety of objects with its legs, feel them with its mouthparts, and retain only those that are suitable. Those that use leaves cut them into pieces of the right size without making wasteful trials of inappropriate objects. As the larvae grow, their cases must be enlarged. Sometimes they take over empty cases abandoned by the original builders. The animal will maintain a particular pattern of case structure, and if an experimenter removes portions of the case, the larva will replace them with pieces of similar size and shape.

Is it possible that even a caddis fly larva has some faint inkling of what it is doing? Most biologists firmly believe that creatures as simple as insect larvae operate entirely by genetically programmed reflexes and cannot possibly think about anything. But if one spells out the genetic programming that would account for all features of caddis fly case construction, it soon becomes a rather imposing list.

Cecropia caterpillars have been studied thoroughly enough to show that the silk cocoons they construct are formed by means of a few relatively simple behavior patterns involving flexion of the body and emission of silk from glands in a certain phase of the movement pattern (Van der Kloot and Williams, 1953). Behavior as complex as the caddis fly larvae's construction of cases and food-catching nets has not been studied in equal detail, but most scientific students of insects argue by analogy that caddis fly cases must result from similar, though necessarily somewhat more involved, reflexes. When a larva cuts a piece of leaf, the dimensions of the piece seem to be related to the size of the animal's head and anterior appendages. This tempts the reductionist to infer that the larva has not made a conscious selection, but that the size of its mouthparts determines the size of the leaf fragment or of the pebble that is picked up. But if we see men using as clubs only sticks of a certain length or weight, we do not deny that they have thought about their selection.

Detailed studies by Hansell (1968) illustrate the degree of selectivity in the construction of caddis fly cases. One species, *Silo pallipes,* begins larval life by constructing a simple tube of sand grains cemented together. After each of its five instars the animal sheds its external chitinous skeleton and grows a larger one. Toward the end of the first instar the *Silo* larva adds to its cylindrical case, composed of half-millimeter particles, two larger sand grains at the sides of the front end. During the second instar it adds two more, larger particles, and through the third, fourth, and fifth instars it follows the same pattern, each time selecting larger grains of sand. By the end of the fifth instar the animal and its case are about ten millimeters long, and the larger, anterior grains of sand are two to five millimeters in size. During all of this growth the larva enlarges its cylindrical case by adding smaller particles. Even this relatively simple creature conforms to definite structural patterns when selecting particles for its case.

Hansell (1972) studied another species, *Lepidostoma hirtum*, that cuts panels from leaves to construct a house with a floor, roof, and two sides, all formed from approximately rectangular pieces of leaf one to two millimeters in size held together by secreted silk. The structure is strengthened by the staggered arrangement of the pieces; each joint between two side plates intersects with the middle of a roof plate, and vice versa. That this staggered arrangement is not wholly accidental was shown when Hansell modified the houses. If he cut away the front end to give the structure a continuous smooth front edge, the larva would cut leaves into different shapes than normal and glue them into place, restoring the staggered arrangement.

According to most biologists, this compensatory behavior can be explained by postulating yet another feature of the genetically programmed behavior pattern. Presumably under natural conditions a house may be damaged, perhaps nibbled by a fish but not totally destroyed. Supposedly, then, for each type of potential damage, the larva's central nervous system is prepared genetically with instructions for repair.

I have discussed the shelter-building behavior of caddis flies in some detail because it poses a very general problem. If we find that an animal constructs a useful and effective structure, how far can we reasonably go in inferring that the animal thinks about what it is doing? A student who suggests such a notion will be firmly corrected by his science teacher and admonished never to mention such an idea again, lest he be judged unscientific. The student is assured that insect larvae operate solely by means of a few fixed reflexes, which, activated sequentially, produce the finished structure. One must postulate that various stimuli activate the reflexes in an appropriate fashion. The larva from whose case a section has been removed does not pick up any old sand grain but feels about until it locates a piece of the right size and shape, or cuts a suitable piece from a bit of leaf.

In a very general sense scientists feel that construction of shelters by animals as simple as caddis fly larvae is similar to the morphological development of complex body parts—except that the actions of appendages or jaws in shelter building are visible, whereas the poorly understood mechanisms that guide anatomical development and cause cells to differentiate into kidneys, legs, gills, and brain are not accessible to our scrutiny. But the two categories are viewed in much the same light, as the results of complex interacting mechanisms, all of which are genetically programmed.

Wasp Nests

Even more remarkable than case construction by caddis fly larvae are the complex structures built by numerous species of termites, wasps, ants, and bees to shelter their eggs, developing young and the adults (Evans, 1963, 1966; Evans and West Eberhard, 1970; Wilson, 1971). The conspicuous paper nests of wasps and hornets result from the combined efforts of numerous females that gather bits of vegetation, chew them with saliva, mix them with silky secretions from specialized glands or with their own feces, and apply the resulting daubs first to a firm substrate and then to the partly constructed nest. The end result is a multichambered nest much larger than the insects themselves.

Among some species of solitary bees and wasps a single female that is ready to lay her eggs goes to great trouble to gather suitable materials from a considerable distance and prepare a place in which the egg has a good chance of developing. This may involve digging a tunnel in the ground or cutting it into solid wood. After eggs have been laid, the tunnel opening may be plugged, and some species select material for the plug that matches the surroundings so that the burrow opening is well concealed. In other species, after the female has laid one or more eggs in the structure, she provides food for the future larvae. The sand wasps (*Ammophila campestris*) studied by the well-known Dutch ethologist G. P. Baerends dig burrows in sandy ground, construct an enlarged chamber at the lower end, and close the opening with small stones, brought from some distance if necessary. Then the wasp captures a caterpillar almost as large as herself, paralyzes it by stinging it in several places, brings it to burrow, opens the entrance, drags the caterpillar down to the nest chamber, and only then lays an egg. The paralyzed caterpillar usually survives, does not decompose, and is eventually consumed by the growing larva (Baerends, 1941, reviewed by Thorpe, 1963).

The burrow sealing and concealment are very effective. Indeed, it is virtually impossible to see any signs of disturbance. Even in the initial stages of burrow digging, the sand grains are thrown some distance away so that they do not create a revealing pile of material. The female wasp locates the hidden burrow entrance by remembering the appearance of surrounding landmarks; if these are altered experimentally, she may be unable to relocate the burrow entrance (van Iersel and van dem Assem, 1964).

Ingenious Nests of Solitary Bees

Nests and shelters take such highly varied forms in the thousands of insect species that I can describe only a few examples here. I have selected these partly on the basis of my intuitive impression that some thinking might facilitate the work of building them.

The solitary mason bees make rather elaborate provisions for storing food and protecting their eggs and growing larvae. Adult female mason bees of the genus *Chalicodoma* gather particles of sand and dirt, which they moisten with saliva and form into small oblong pellets. The female then carries each pellet to the surface of a large rock and cements them together into a roughly cylindrical cell, open at the top. She lays one egg in each cell and places liquid honey close to the egg. She then closes the top of the cell with more sand grains cemented with saliva. Then she applies dusty particles to the outside of a group of cells, making them look almost exactly like the surface of the rock to which they are attached. The mixture of fine particles and saliva becomes almost as hard as cement. When the larva grows and changes first into a pupa and finally into an adult bee, it must exert considerable effort to chew its way out of the cell (Frisch, 1974).

Another type of mason bee, *Osmia bicolor,* uses a different procedure to provide for its progeny. The adult female seeks out an empty snail shell and deposits eggs and food in the inner, narrow portions of the spiral chamber that once protected the snail. Some of the food is a semisolid containing pollen and known as "bee bread." After she has deposited a number of eggs and a quantity of bee bread, she fills the middle portion of the tapering spiral cavity with chewed-up pieces of leaf and, nearer the outside, enough small pebbles to constitute a fairly rigid wall. The accumulation of leaf fragments and pebbles is not airtight and, even after a second wall of leaf pulp is added outside of the pebbles, enough air can enter to provide oxygen for the growing larvae. After all this has been completed, the *Osmia* mother collects dry stalks of grass, tiny twigs, or pine needles, which she piles over the snail shell in a large, irregular, porous dome that hides the shell. Certain species of "cuckoo bees," instead of building their own structures, lay eggs in the partly completed nests or partly filled snail shells prepared by bees such as *Chalicodoma* or *Osmia* (Frisch, 1974).

What thoughts might these hard-working mother bees experience? They can scarcely think about the next generation, because the mothers die long before their offspring emerge. But they might think about the results of their elaborate nest building efforts in the short-term future. When a female *Osmia* has laid eggs and provided bee bread in the inner portion of the snail shell, she might think, "Now I want to close the rest of this cavity." When she brings pebbles or leaf pieces to the shell, she could be thinking about leaving air spaces while closing up the cavity, even though she cannot understand the adaptive value of the barriers and the air passages. When a female *Osmia* is gathering material to cover her snail shell, she may not understand the long-term advantages of hiding it, but she may have simple thoughts about her immediate objective. Perhaps she remembers the nest from which she emerged, or perhaps she anticipates the filled snail shell.

Leaf-Cutter Ants

The shelters of social insects are built through the cooperative efforts of many individuals. Termites, ants, wasps, and bees build highly varied structures, both underground and above ground, in an enormous range of size and complexity. All of these are fascinating, but in the interests of brevity, I will consider only some of the larger ant nests, because they are produced by one of the most elaborate social organizations found outside of our own species. In this leap from mason bees to the most advanced ant societies I am ignoring a great many intriguing social insects that build almost equally complex shelters (Wheeler, 1910, 1928; Wilson, 1971).

One of the most successful animals in the world in terms of numbers of well-nourished individuals and impact on the environment are the leaf-cutter ants of the New World tropics (the genus *Atta* and its close relatives). They live in huge underground colonies consisting of many chambers connected by tunnels. Each colony is made up of a single queen and thousands of nonreproductive workers. Reproductive males and virgin queens are produced only after the colony has grown substantially. The worker ants are adapted both in anatomy and behavior for different functions, such as caring for eggs, larvae, and pupae; gathering food; or defending the colony against predators or intruding insects of other species. I will sketch very briefly only a few of the many operations carried out by the workers. From the colony's entrance on the forest floor the food gatherers move out in such numbers that they quickly wear down the vegetation and form beaten paths over the ground. They climb small plants or tall trees, cutting pieces of leaf, roughly the size of their own bodies, which they carry back to the colony. Hundreds of them can be seen walking methodically along their trails, each carrying a tiny green fragment of leaf. They sometimes attack flower beds, and the dismayed gardener may discover a busy trail of ants, each carrying a 5-millimeter fragment of colorful flower petals to the underground tunnel system (Weber, 1972).

Inside the colony the workers carry the leaves or flower petals into special chambers, some as large as a meter in diameter, containing masses of fungus that grows rapidly, nourished by the leaf fragments, and provides food for the leaf-cutter ants. To help the fungus mass grow, worker ants chew the leaf fragments, mix them with saliva, and add a few of their own feces. The metabolic processes within the fungal cells convert the indigestible cellulose in the leaves into sugars that the ants can digest.

Occasionally, a number of winged reproductive males and females are produced. When they leave the colony, they appear as a conspicuous swarm. Each female takes a tiny bit of fungus, which she places in a specialized pocket inside her mouth. After mating, she regurgitates this material in her newly dug burrow, where it forms the beginning of a new fungus garden, which is absolutely essential for the survival of the new colony (Weber, 1972). Thus these ants, like those of several other species, are in a very real sense carrying out a form of agriculture; they collect a particular food plant, bring it into the colony, and help it grow.

What thoughts and feelings may be experienced by the ants that carry out such a specialized form of agriculture? It is generally assumed that genetic programming rather than learning directs all their patterns of behavior—the digging of underground chambers, which may extend 3 to 5 meters below the ground, the collection of huge amounts of "fungus food," the tending of the fungus gardens, and the removal of other types of fungus, as well as caring for the eggs and larvae. It would be very difficult to rear leaf-cutter workers in isolation to ascertain whether their behavior is influenced at all by the activities of all their many sisters. But when a new colony is started, it is clear that the first workers to develop cannot learn anything about the behavior they will shortly carry out by observing other ants.

Can we reasonably infer from the varied, effective, and highly integrated behavior of leaf-cutter ants that they might think consciously about burrow construction, leaf gathering, fungus gardening, or other specialized activities? As in other instances, prevailing biological opinion is vehemently negative. Yet the principle of adaptive economy, outlined in Chapter 2, may appropriately be called upon in this instance. The workers of leaf-cutter ants are tiny creatures, and their entire central nervous system is less than a millimeter in diameter. Even such a miniature brain contains many thousands of neurons, but ants must do many other things besides gathering leaves and tending fungus gardens. Can the genetic instructions stored in such a diminutive central nervous system prescribe all of the detailed motor actions carried out by one of these ants? Or is it more plausible to suppose that their DNA programs the development of simple generalizations such as "Search for juicy green leaves" or "Nibble away bits of fungus that do not smell right," rather than specifying every flexion and extension of all six appendages?

I should like to reiterate how little of the wealth of insect social behavior I have touched upon in this brief discussion. Many kinds of colonial ants carry out equally elaborate and specialized forms of behavior. Some gather and care for other species of insects. Certain ants feed on the sugary feces exuded by aphids and scale insects, which adhere to plants and suck their sap. In some species the ants actively protect their "domesticated" livestock. They may even build shelters around them, and the relationship becomes a true symbiosis that is extremely beneficial, even essential for both species (Wilson, 1971). Nest structures are also highly varied and adapted to the needs of the particular species. The leaf-cutters suffice to illustrate my general theme, but one other specialized social ant also deserves consideration.

Weaver Ants

African weaver ants (*Oecophylla longinoda*) live primarily in trees, but rather than hollowing out a burrow or other cavity, they construct their nests by joining together the edges of leaves with sticky silk. As in most other ants, the nest is constructed by nonreproductive female workers. The first problem they must solve is to bend the leaves away from their normal positions so that the edges meet to form the walls of a suitable cavity. In some cases the edges of the leaves are already close enough that an ant can grasp one leaf with her rear legs and the other with her jaws. Then by bending her whole body and flexing her legs she can pull toether the edges of the two leaves. The individual ants, however, are far smaller than the leaves they use, so numerous workers line up along the edges of the two leaves and all pull their edges together. Even this degree of cooperation is not enough, because the edges of the leaves are often separated by far more than the ant's fully extended body length. They solve this difficulty by forming chains of ants. One seizes the edge of a leaf in her jaws, another grasps the abdomen of the first ant, a third holds the abdomen of the second, and so on, until at the end of a chain of several individuals, one seizes the other leaf with its hind legs. Numerous ants pull and bend the leaves until they form a tight enclosure.

After the leaves are pulled together, the edges are joined with sticky silk. The adult workers do not secrete this silk; instead, they bring larvae of the appropriate age from the nest and hold them first against one leaf edge and then the other. Thus the larvae, which are of course younger sisters, are used as a sort of tool in nest construction.

Within this enclosure the queen lays her eggs, larvae grow and are fed by workers bringing in food, mostly fragments of other animals they have killed or scavenged, and all of the many maintenance activities of an ant colony go an. One leaf nest is seldom large enough to hold a growing colony, so while the queen remains in one nest, workers construct others nearby in the same tree. There is then a great deal of traffic back and forth between the nests, which constitute a single colony. In some cases these colonies contain dozens of nests extending over several trees. The workers travel back and forth carrying eggs from the queen to the other nests and feeding the developing young wherever they may be.

All this behavior of weaver ants would be facilitated by some simple conscious thinking. When two leaves must be pulled together to start nest construction, the workers might think something like "Let's pull those two leaves closer." While genetic instructions might specify all of the motor actions necessary to pull together leaves of different sizes, shapes, and separations it might be more economical to transmit from one generation to the next the mental image or concept of two leaves close together, edge to edge, enclosing a cavity. This would suffice, if the workers can judge what to do to change the separate leaves moving freely into a nest wall stuck together by the edges. Similar speculations about the use of larvae as sources of sticky material are equally plausible (or, many will say, implausible). Could these weaver ants pulling two leaves together think "Those larvae put out sticky stuff that would help hold these leaves together"?

It is probably significant that these same specialized weaver ants also communicate with each other about activities away from the nest that are important to the colony as a whole. I will return to this communication behavior in Chapter 9.

Bird Nests

Birds' nests are far more familiar than those of insects. But when allowance is made for the difference in size of the builders, most birds' nests are rather small and simple compared to the elaborate nests of wasps, bees, and ants. This fact is somewhat disturbing to those who remain convinced that only those animals most closely related to ourselves can exhibit versatile and possibly thoughtful behavior. Some bird nests, such as those of orioles, which are elaborate, covered structures with long entrance tubes, are impressive. More commonly, the nest is basically cup shaped and open to the elements except when occupied by an incubating parent. They are often well concealed, and during both the construction and the incubation of eggs, parent birds try to arrive and depart inconspicuously. The adaptive significance of concealment and inconspicuous behavior in the vicinity of the nest are obvious, since eggs and nestlings are easy targets for predatory mammals or other birds, which can find a nest easily if it is visible or if the parents are conspicuous when coming or going.

Most bird nests are apparently built primarily on the basis of genetic instructions, and such instinctive behavior is held to preclude any conscious thinking. Nevertheless, nest building is anything but a stereotyped and fixed sequence of behavior patterns (Thorpe, 1963). While many of the individual actions are repeated consistently from one occasion to the next, others vary greatly, and the sequence is almost never precisely the same. Rather, it seems to be adapted to the task at hand. If a nest is damaged, the bird ordinarily will repair it, although it may abandon the whole enterprise. Furthermore, when an experimenter takes away or supplies materials they would ordinarily have to gather, the birds behave sensibly. For example, if the soft feathers used for lining a nest are removed, the bird replaces them with more feathers, sometimes from its own body. But the same bird will readily use material supplied by the experimenter and refrain from gathering more of the same.

The complex nests built by the males of several species of African weaverbirds have been described in detail by Collias and Collias (1964) and by Crook (1964). The thoroughly studied common village weaverbird (*Textor cucullatus*) weaves a gourd-shaped nest consisting of a cuplike egg chamber, lined with feathers or soft grass heads and surrounded by a shell about 15 centimeters in diameter. Strips of grass about 30 centimeters long are inter-woven to form the walls, but the ceiling is formed from shorter pieces of grass more tightly interwoven so that they shed rain more effectively. Next to the nest cup is a circular hole in the bottom of the nest through which the birds enter from below. As described by Collias and Collias (1964), the male first weaves long thin strips of grass into a ring attached to suitable branching twigs. Then the ring is extended to produce a roughly spherical structure. The individual motions involved are fairly complex and are re-peated over and over. The bird grasps one end of a strip, thrusts it through an opening in the nest material, lets go, seizes the end of the strip on the other side, pulls it, and thrusts it through another hole, each time alternating the direction in which the strip is bent. But this simple description fails to tell the whole story. For one thing, the bird is always weaving a curving surface, although with the same motions he could produce a solid ball of woven material or many other shapes. After the nest is nearly completed and the male has made many courtship displays, a female may accept the nest. If not, the male usually tears it apart, builds another nest and continues courtship behavior.

A downward-facing entrance tube to the nest is constructed around the entrance hole, but only after the female has moved in. The male weaves strips of leaf or grass to produce a smooth edge at the entrance to the tube. If an experimenter roughens this edge by pulling out some pieces, the male adds more material before making the edge smooth, and in this way he can be induced to build an abnormally long entrance tube. When a partly completed nest is damaged, the male usually reconstructs the missing portion. Although Crook (1964) reported that wild weaverbirds whose nests were damaged often repeated earlier phases of nest construction unnecessarily, the captive weaverbirds studied by Collias and Collias (1962) did not repeat all the sequences when confronted with a damaged nest; they did only what was necessary to repair the damage. The major exceptions to this rule are the construction of lengthened entrance tubes after the normal edge has been roughened and the common reaction of either abandoning a nest that has been damaged by the experimenter or tearing it apart and using the materials to construct a new one.

In general, birds are more flexible and sensible than insects in repairing damaged nests unless they abandon the nest. In terms of evolutionary adaptation, abandonment is probably an effective strategy, since a nest that has been damaged once may well suffer similar damage again, and the bird's chances of rearing young may be enhanced by a new effort.

This is not to say that birds never do foolish things in the course of nest building. Von Frisch (1974) describes how blackbirds will occasionally start building many nests in some artificial structure that has many similar-looking cavities. The birds apparently become confused as to just where the nest is to be and never succeed in completing any one nest. As in so many cases of this kind, we tend to infer a total lack of thinking when animals do something foolish and wasteful of effort. But we do not apply the same standard to members of our own species, and we never infer a total absence of thinking when people behave with comparable foolishness. It is important to realize that to postulate that an animal may engage in conscious thinking is by no means the same as to say that it is infinitely wise and clever. Human thinking is often misguided, and there is no reason to suppose that animal thinking always corresponds perfectly to external reality. People's thoughts may be inaccurate or quite different from what other people see as correct and sensible. But error is not the same as absence of thought.

Events in our own lives show that stupidity does not preclude consciousness. If our usually dependable automobile fails to start, in our impatience many of us do something totally irrational, such as kicking the tires or swearing at the machine. We know perfectly well that such displaced aggression will not start the car, but we are probably thinking "Why won't the damned bus start this morning?" or "What will happen when I am late at the office?" Or "I should have bought a new battery last fall!"

This leads us to another challenging question about nest building and other complex reproductive behavior of birds. When they are at an early stage of nest building, do they have any concept of the finished product they are working to achieve? We might go one step further and ask whether a bird beginning to build a nest has any idea of the eggs and young that will soon occupy the nest. Such speculation may be more plausible in species where the females do the nest building, for they will lay the eggs and feed the young. Many will feel that it is outrageously far-fetched to suggest that a female bird might think about eggs and young as she begins to build a nest, but there might be some advantages to such thoughts. Even a simple concept of the function the nest will serve could help birds construct it appropriately. But again we are stymied by our inability to gather convincing evidence.

Thoughtless Robots?

Despite the versatility of this behavior, scientists generally believe that it all results from genetically programmed instincts and that it cannot therefore involve any conscious thought. One reason given in support of this equating of instinctive with unconscious is that under some conditions of human intervention the behavior of insects is inappropriate and ineffective. If an experimenter removes some of the caterpillars that a wasp has brought to provision her nest, or adds additional ones, the wasp will often continue to bring in the same number of caterpillars as she would have done if left undisturbed. If a paralyzed caterpillar is moved after the wasp has laid it near the burrow entrance while she reopens the hole, she often hunts for the missing prey and drags it back to the entrance. But rather than carrying the caterpillar down the shaft directly, she repeats the entire behavior pattern of depositing the caterpillar near the shaft and digging, even though the burrow is already open. This sequence of inappropriate behavior may be repeated many times.

In other cases, insects that have constructed a nest fail to alter their behavior if an opening is made in the nest so that the eggs fall out. This has led to the conclusion that the insect cannot possibly be thinking consciously. We can immediately see what she ought to do to accomplish the general objective, and when she fails to do this and instead repeats what would ordinarily be appropriate behavior, we conclude that this animal is a completely mindless machine (see, for instance, Gould and Gould, 1982).

It is worthwhile to examine our train of reasoning with some care. Is it the consistency of the wasp's behavior from one occasion to the next that makes it seem mechanical? If so, we must recognize that the details of the behavior do vary far more than would be likely in a mechanical wasp, if we could build one. Do we deny conscious thinking primarily because the wasp fails to show the sort of insight we expect of ourselves? Is this justified? If a child of, say, six does something foolish when adults can see how he could solve a problem that baffles him, we do not argue that he is not thinking consciously. Could we be committing a comparable fallacy in the case of the wasp?

Wasps and many other arthropods often do adjust their actions to varying circumstances, but these cases are seldom emphasized in discussions of instinctive behavior. For example, W. S. Bristowe (1976) describes how orb-weaving spiders sometimes vary their stereotyped behavior in dealing with small insects caught in their webs. If an experimenter holds a struggling fly with forceps close to such a spider, she omits the earlier stages of normal behavior (running along the web to reach the fly) and bites it immediately. If the fly is already dead, she wraps it in silk without biting it first. In constructing their elaborate webs, spiders are often said to follow a rigid series of behavior patterns which are presumably instinctive since a female spinning her first web does so almost perfectly. But she will make some alterations in structure when the surrounding vegetation or the space to be spanned is irregular. Bristowe describes how a spider whose web is ordinarily symmetrical builds a highly asymmetrical web when the opening between leaves makes such a shape appropriate. At the web's hub from which strands of silk radiate out to the surrounding vegetation, the spider ordinarily leaves a hole so she can quickly move from one side of the web to the other when an insect strikes it. In one web this hole, instead of being at the center, was close to one edge of the opening between the leaves of a lilac bush, and the strands formed a semicircle instead of a circle.

Many ethologists dismiss variability in structures such as spider webs as meaningless "noise" in a basically invariant system and deny that a spider could consciously adjust the structure of her web according to the shape of the available opening. But the end results are so efficiently adapted to their function of catching small flying insects that it seems possible that spiders anticipate the likely results of their web spinning. This inference will be strongly disputed by most entomologists and ethologists, however, because they are so accustomed to thinking of insects as genetically pro- grammed robots. But we know far too little to assume that ge- netically programmed behavior must necessarily preclude conscious thinking.

Wasps are able to alter their behavior to deal with conditions that vary within a certain range. For example, if a wasp is digging a burrow in loose soil and particles fall in, she removes them (Baerends, 1941; Thorpe, 1963). The wasp's specific motions de- pend upon when and where the particles fall and no removal motions occur if nothing falls into the partly completed burrow. But when a human experimenter changes the circumstances to something well outside the natural range, for instance by moving a paralyzed cricket which a wasp has laid beside the entrance of her burrow while she inspects the burrow prior to carrying the cricket inside, we are disappointed when the wasp retrieves the cricket but then needlessly repeats the burrow inspection (Gould and Gould, 1982). The wasp fails to behave in what *we* can easily see would be the most efficient manner under the circumstances. But a lack of versatility in the face of wholly unprecedented cir- cumstances does not necessarily mean that the behavior is un- conscious.

Our competent and well-informed leaders may eventually come to be viewed in a similar light. Our species has had a long history of warfare and preparation for violent aggression against other human groups. It is widely believed, with good reason, that if potential adversaries know that the enemy has a large supply of potent weapons, they are discouraged from attacking. But our species has now learned to produce weapons which, if used on any substantial scale, would almost certainly destroy not only all potential enemies but ourselves as well. Yet we continue to repeat the same patterns of preparation for warfare. Fortunately, we have managed to avoid nuclear warfare so far, and we can all fervently hope that we will continue to do so. Otherwise, extra- galactic palaeoethologists may someday reconstruct our terminal history from a viewpoint similar to that of the ethologist who watches a wasp repeat sequences of behavior that are ordinarily effective when the changed conditions have rendered them mal-adaptive.

In interpreting the complex nest- and shelter-building activities of insects, we must be on our guard against jumping to sweeping conclusions on the basis of limited evidence that happens to fit with our prior convictions. Although insects often act stupidly and ineffectively when their nests are experimentally damaged, others, such as termites, do repair damaged nests. Even caddis fly larvae repair holes opened in their cases. Another example is provided by potter wasps, which build pot-shaped structures from daubs of mud gathered at some distance. They place provisions for a future larva in these containers, and when it is nearly full, they lay an egg. When a hole is made in the bottom or side of the pot and some of the provisions removed, the wasp sometimes notices the hole and repairs it (reviewed by Thorpe, 1963). If we conclude that a species of wasp cannot be thinking when it behaves stupidly, must we switch to the opposite opinion when we find that a different species behaves sensibly in a similar situation? There has been a strong tendency to emphasize the stupid cases and ignore insects that do act sensibly.

Another argument often made is that insects' elaborate construction activities cannot entail conscious thinking because the insect will not see the eventual result. Wasps build elaborate nests, provide food for future larvae, lay eggs, and carefully close the whole structure, concealing its entrance. But the adult female insect that does all this dies before the larvae take advantage of her work. It is thus argued that she can have no conception of the eventual result of her elaborate activities. Since the mother was once a larva herself, perhaps she would remember enough to construct something similar. But drastic changes take place in the insect's body, including the central nervous system, during metamorphosis from the larval stage to the adult wasp that does the nest building. It seems highly unlikely that any detailed memories could be carried through this process.

Nevertheless we should ask whether it is necessary to understand the ultimate results of a behavior pattern to have any conscious thoughts about it while it is under way. The thoughts and feelings of a caddis fly larva or a nest-building wasp may be limited to the immediate situation rather than its long-term consequences. The wasp may want to build a burrow, she may plan to bring the paralyzed caterpillar into the nesting chamber, and she could enjoy laying eggs—all without the slightest understanding that this will result in a larva of her species. The conviction that an understanding of long-term results is a necessary condition for conscious thinking may arise from our need to find reasons for believing that animals are unthinking machines. Many kinds of human behavior are carried out with some thought about the immediate circumstances but none whatever concerning long-term results. Few of us think about our descendants in A.D. 3000, although our behavior is crucial to their future existence. And the lighting of cigarettes and enjoyment of smoking are seldom accompanied by detailed contemplation of the terminal agonies of lung cancer or heart disease.

Inherited Templates

The suggestion that instinctive, genetically programmed behavior may be adapted to changing circumstances conflicts sharply with contemporary scientific thinking, although it would not have disturbed Charles Darwin or G. J. Romanes (1884). But a helpful bridge may be provided by the concept of genetically programmed perceptual templates. Rather than a precise replica of something in the outside world, a template is a representation in the brain coded in ways we only dimly understand, as discussed by Roitblat (1982). It could be any sort of pattern within the animal's brain that it strives to match. To elaborate a bit upon this plausible but sketchy notion, let us suppose that an animal's DNA leads to the development within its brain of a mechanism that responds selectively to a particular pattern of simple or complex sensory input, and that the animal seeks to alter the sensory input from its environment to match the neural template.

Something of this sort seems to underlie at least in part the ability of songbirds to produce their species-specific songs. If the young of some songbird species are kept in isolation and hear recordings of many different sounds, their adult songs resemble the recordings. But careful experiments have demonstrated that they learn some patterns more easily and more accurately than others; the most easily learned patterns resemble the characteristic song pattern of the species. Thus, while learning can lead to quite abnormal songs, some song types are more likely to be learned under natural conditions (Marler, 1976, 1977, 1978).

Some sort of neural template conforming to the sensory input received from nests, burrows, prey-catching devices, or other structures might develop in an animal's brain, without learning or individual experience. If constructing artifacts does involve producing something to match the template, it is not surprising that the behavior may vary and be adapted to the immediate circumstance. Thus, if caddis flies have a neural template corresponding to the sensations received from a certain type of case, we might postulate that after placing grains of sand or bits of leaf around their bodies in a pattern that conforms to the template, this portion of the structure will be retained. On the other hand, if the initial results are discordant with the template, the pebble or leaf fragment will be replaced by something that matches it better. Appealing as this hypothesis is, it would require elaboration to explain all aspects of artifact construction, even in caddis fly larvac, for they often appear to select or prepare objects of the right size and shape rather than trying any available particles and discarding those that are unsuitable. The neural templatc would have to include the "feel" of suitable particles as well as the feel of a finished case. The concept of inherited neural templates could help to explain the complex and adaptive behavior exhibited by many kinds of animals even when they have no opportunity to learn what to construct.

This concept can stimulate our thinking and suggest fruitful observations and experiments, even though it is still only a sketchy sort of outline that needs to be fleshed out by a great deal of specific factual information before it could become truly convincing.

The general principle of neural templates can also help us in one other way. If the brain of an animal develops in such a way as to produce neural templates, and if the animal perceives the nature of the template, it might then know, so to speak, what it is trying to produce. If this notion seems doubly fanciful, consider how it might operate at a very simple level. One basic neural or perceptual template might be the sensation of being in a snug, dark cavity that is safe from most hazards of predation. The template might consist simply in feeling, by a variety of sense organs, cosily surrounded by the walls of a cavity, coupled with the visual perception of dim light. If animals are aware of such a "goal template," they might be capable of consciously guiding their behavior toward matching it. In other words, the animal's behavior might result from a conscious attempt to realize—literally, to make real—the perceptual template.

One neural template in caddis flies might involve the tactile sensation of one's sensory spines being in close contact with firm walls over most of the body surface. The larva could make its sensory stimulation conform to that template by cementing together particles to form a cylindrical tube. Such a perceptual template would allow the animal to think in terms of the completed structure when it was in various stages of making the case correspond to the pattern. Since the cases have some properties that are specific to the particular genus of caddis fly, the neural templates would have to include these attributes. For those that cut square pieces of leaf and cement them together with staggered joints, the neural template must involve the tactile sensations produced by such plates and joints.

Explaining instinctive behavior in terms of conscious efforts to match neural templates may be more parsimonious than postulating a complete set of specifications for motor actions that will produce the characteristic structure under all probable conditions. Conscious efforts to match a template may be more economical and efficient. It is always dangerous for biologists to assume that only one of two or more types of explanation must apply universally. Probably the various sorts of instinctive behavior are based on a number of combinations and permutations of neural templates and genetically programmed motor behavior. Neural templates may also undergo changes through learning and perhaps through nonspecific environmental influences or maturation of the animal's nervous system. They may well be quite flexible and merge into searching images of the kind discussed in Chapter 3. And it is not necessary to suppose that animals are consciously aware of all their neural templates; perhaps only a few are important enough that the animal thinks consciously about them and considers alternative ways of realizing them.

Even though these speculations are highly tentative, they do provide a plausible basis for considering how conscious thinking might be combined with instinctive behavior carried out without any opportunity for learning. Most of an animal's thoughts and subjective sensations are probably confined to the immediate situation rather than ultimate results. A spider wanting to build a web may think about how to manipulate her spinnerets to produce different types of silken thread, but she may not be able to imagine that the structure she is constructing will help her catch insects. Or she might be able to anticipate, when a web is partly constructed, that the completed web might catch the fly that just flew through one of the remaining gaps. When a female wasp digs a burrow, provisions a nest, or makes other arrangements for an egg she has not yet laid, in a future she will not live to experience, it is unreasonable to imagine that she thinks about her eventual progeny, primarily because there is no way for information about the progeny to reach her central nervous system. But this inability to know the long-term results of her behavior in no way precludes conscious thinking about what she is doing, for she has abundant opportunity to perceive what she does and to observe the immediate results.

6 Tools and Engineering

For a long time it was widely argued that only our species uses tools, or at least that animals cannot prepare or construct tools. But patient students of animal behavior have turned up several examples of tool use. In a few cases the animal modifies something available to make a better tool. Unflattering as it may seem for our human conceits, such tool use is not concentrated in our nearest relatives, the great apes and the monkeys. It is found in a variety of animals, including several invertebrates. Tool-using species are a small minority, and in some it is confined to a few individuals, or to certain populations. Animals that use tools do so only when it is important to them, usually when food can be obtained only with the aid of some implement.

The use of tools by animals has recently been reviewed in a thoughtful and authoritative book, *Animal Tool Behavior*, by Benjamin Beck (1980). Since this survey is so recent and so thorough, I shall draw upon it extensively; except where other sources are cited, the examples discussed in this chapter are from Beck, where they are described in greater detail. For our purpose, the important consideration is whether manipulation of objects or the use of tools provides evidence that the animals are intentionally thinking about what they are doing. Several of the examples of tool use and the construction and maintenance of quite elaborate artifacts discussed in this chapter seem more likely than those reviewed in Chapter 5 to involve anticipation of results the animal hopes to achieve in the near or distant future.

Ant-Lions and Other Invertebrates

Two distantly related groups of insect larvae excavate funnel-shaped holes in loose soil and wait, buried near the bottom of the cavity, for other small animals to fall in, which they then capture and eat. The best known of these pitfall-constructing insects are the ant-lions, which metamorphose when mature into flies of the order Neuroptera. The larvae of certain true flies (Diptera), called worm-lions, have similar habits; like the ant-lions, they construct pitfalls to capture prey and also throw sand grains at escaping prey. The conventional viewpoint insists that insect larvae could not possibly be aware of anything. But digging pitfalls and throwing sand grains certainly involves adapting behavior to the local circumstances. The ant-lion selects soft dry soil for the pit, and throws sand not at random but only in the direction of the prey.

Another significant example is provided by hermit crabs, widely distributed and locally abundant small crustaceans that occupy the empty shells of dead snails. The rear portion of a hermit crab's body is shaped to fit inside a snail shell, without which it is very vulnerable. As it grows, it must find increasingly larger shells. A crab moves into an empty shell only after inspecting it and feeling it with the anterior appendages, apparently to find out whether the shell is suitable. Hermit crabs often fight over a shell, and in one species they congregate where predators are feeding on snails and wait for empty shells to be dropped.

Sea anemones have nematocysts that discharge a stinging and irritating mobile projectile when something touches them. The marine crab *Melia tessellata* carefully detaches small anemones from the substrate, holds one in each of its two anterior appendages, and when threatened directs the anemone toward the intruder (Duerden, 1905; Thorpe, 1963). Some hermit crabs place anemones on their shells, and thus gain some protection from predators (Ross, 1971). It is an indication of ethologists' lack of interest in behavior suggesting conscious thinking that the crabs' use of anemones has received virtually no serious investigation in the almost eighty years since it was described by Duerden. It is not even mentioned in Maier and Schneirla's *Principles of An-*

imal Psychology (1935) or in most monographs and textbooks devoted to ethology. If a primate was found to behave as ingeniously as these crabs or ant-lions, we might credit them with intentional planning. If we adamantly deny any conscious intention to these crustaceans and insect larvae, we must recognize that we do so for reasons other than the nature of their behavior.

A few species of ants provide the clearest examples of tool use by small and relatively simple animals. Worker ants of the genus *Aphaenogaster* collect soft and semiliquid foods such as fruit pulp, honey from the stores of other insects, or the body fluids of animals they have killed (Fellers and Fellers, 1976). The workers must transport these semiliquid foods back to their nest rapidly before other animals eat it. To do so, the ants often pick up bits of leaf, wood, or even mud and place them in the liquid long enough for appreciable amounts to be absorbed, then carry back the wetted spongelike object to the colony. By using this sort of tool, a worker can carry back some ten times as much food as it could carry in its stomach.

Another example of tool use by insects is that of the digger wasps, whose behavior was discussed in Chapter 5. In the course of their elaborate burrow building, certain of these wasps tamp down the soft dirt with a small pebble held between their mandibles. As pointed out by Thorpe (1963), the tamping motion is somewhat similar to the motions they use when they lift particles of soil and drop them into the burrows that are being filled up. Nevertheless, there is some differentiation of the motor behavior, and the tamping occurs only toward the end of the whole filling process.

Tool Use by Birds

Several species of birds have been observed to use simple tools (Chisholm, 1954, 1971, 1972). Stewart W. Janes (1976) observed nesting ravens make an enterprising use of rocks. He had been closely observing ten raven nests in Oregon, eight of which were near the top of rocky cliffs. At one of these nests two ravens flew in and out of a vertical crack that extended from top to bottom of a twenty-meter cliff. Janes and a companion climbed up the crevice and inspected the six nearly fledged nestlings. As they started down, both parents flew at them repeatedly, calling loudly, then landed at the top of the cliff, still calling. One of the ravens then picked up small rocks in its bill and dropped them at the human intruders. Several of the rocks showed markings where they had been partly buried in the soil, so the birds presumably had pried them loose. Only seven rocks were dropped, but the raven seemed to be seeking other loose ones and apparently stopped only because no more suitable rocks were available.

While many birds make vigorous efforts to defend their nests and young from intruders, often flying at people who come too close, regurgitating or defecating on them, and occasionally striking them with their bills, rock throwing is most unusual. Nor do ravens pry out rocks and drop them in other situations. It is difficult to avoid the inference that this quite intelligent and adaptable bird was anxious to chase the human intruders away from its nest and decided that dropping rocks might be effective.

Although birds have not often been observed to pick up and drop rocks, some crows and herring gulls have developed the habit of breaking shellfish by dropping them on hard surfaces, as described in Chapter 3. No one has observed the same bird performing these two types of dropping behavior, so it seems unlikely that the ravens were simply shifting from one type of dropping behavior to another. Nests of ravens and other birds have often been studied under similar conditions, and rock dropping to deter intruders would probably have been reported by ornithologists if it had occurred. It thus seems likely that the ravens observed by Janes had hit upon this novel defensive strategy independently. It would be fascinating to know whether they had used it on other occasions and, if so, with what success.

In any given species, especially enterprising behavior patterns are probably rather isolated. If individuals hit upon a certain tactic and employ it repeatedly and effectively, it is not ordinarily imitated by other members of the species. Independent discoveries that do not spread through a population are not of interest to evolutionary biologists who concentrate on phenomena in large populations, but such inventive behavior may be all the stronger evidence of independent thinking, as emphasized by Dennett (1983). When intruders threaten a raven's nest and do not desist in response to loud vocalizations, threatening approaches, and other tactics, the bird may think up the new tactic of dropping some hard and possibly dangerous object. It would be interesting to know whether these birds have observed the effects of dropping stones or whether they reason that a falling stone might deter the intruder. Again, we do not yet have any way of answering such questions.

Another well-studied case of birds using stones as simple tools is the breaking of large eggs by certain vultures in East Africa. These birds often eat the eggs of ground-nesting birds, but ostrich eggs are too large and too tough to break by simple pecking. The vultures sometimes break smaller eggs by throwing them 60 to 100 cm. When confronted with a large ostrich egg, a vulture may pick up a stone in its bill and hammer at the shell until it breaks, or it may throw stones at the egg. They miss about half the time, but they will persist if not disturbed until the shell breaks (van Lawick-Goodall, 1970).

One of the best-studied cases of birds preparing and using tools involves the Darwin's finches of the Galapagos Islands discussed in Chapter 3. On most of these islands the finches feed on insects, which they scratch out from crevices. Some species have been observed to use tools to help them (Lack, 1947; Bowman, 1961; Millikan and Bowman, 1967). The bird selects a cactus spine or small twig, which it holds in its bill, sometimes first modifying it by shortening or breaking off protruding portions. This artificially lengthened bill is then probed into crevices, and the insects are impaled, scraped out, or simply battered about until they are forced out. The bird then drops the twig and eats the insect; but sometimes it holds the twig under its toes and uses it again. When this behavior was observed in captive Galapagos finches, some enterprising birds retained their tools for as long as two minutes. Millikan and Bowman found that finch species that had not been observed to use tools under natural conditions did so in captivity after living for a year next to cages occupied by accomplished probers. Several distantly related species have been occasionally observed to use tools in a similar manner to reach otherwise inaccessible food, but such behavior seems to be relatively rare.

Jones and Kamil (1973) observed that one of eight captive blue jays learned to use short sticks to reach food that was not otherwise obtainable. Five of the remaining seven jays later began to use similar tools, almost certainly by observational learning. Jays are ingeniously versatile birds, and similar tool use has been observed under natural conditions by Gayou (1982) in a family of green jays in Texas. Much like the Galapagos finches, these birds picked up small twigs in the bill and used them to poke about under loose bark and to dislodge insects. Sometimes they held on to the twigs to use again in the near future. A young jay was observed trying unsuccessfully to use a twig, as though it was imitating one of its parents but had not yet learned how to do so effectively. Tool use was observed in only two out of fourteen green jays observed, and it entailed only about 5 percent of the feeding activities. It may well have been another case of isolated invention, in this case by one family.

Baiting Prey

Beck quotes a fascinating observation by Lovell (1958), who watched a green heron pick up small pieces of bread, drop them onto the water, and then capture fish that were attracted to this bait. The heron seemed to place the bait, which it had brought from some distance, at a place where it had seen fish, and it retrieved the bait when it drifted away. As in other similar cases, we do not know what experience had preceded the bird's observed use of bait. Such behavior, which seems to be an isolated "invention" by one individual, has been reported in a few other birds. An Australian ornithologist, Greg J. Roberts (1982), reported that he observed a black kite pick up a scrap of bread from a campground, carry it over a small river, and drop it into the water. The kite then landed on an overhanging branch, and when crayfish were attracted to the bread, the kite repeatedly tried unsuccessfully to catch them in its talons. Black kites are known to be scavengers, but they do not ordinarily capture aquatic animals. Thus trying to catch crayfish in shallow water and luring them with bait represented a dual extension of the kite's normal feeding habits. Since this occurred at a campground, the kite might have seen crayfish attracted to bits of food dropped by people, but it went one step further and dropped its own bait. Or perhaps the kite had previously scavenged food from the campground and had accidentally dropped some into the river. If it saw that crayfish were attracted to the scraps, it may have "put two and two together." All this is merely conjecture, pending more observations by interested ethologists, but it does suggest that the kite was capable of creatively modifying its behavior.

Elizabeth McMahan of the University of North Carolina has discovered that in tropical rain forests one species of assassin bug, a predatory insect, uses two effective tricks to capture the workers of termite colonies (McMahan, 1982). The bug glues small bits of the outer layers of a termite nest to its head, back, and sides. Then it stands near an opening to the termite colony. The bits of termite nest on the assassin bug apparently smell and perhaps feel familiar to the termites, so no alarm signals are emitted, which otherwise would attract the well-armed members of the soldier caste that attack intruders. Although the assassin bug's actions often attract soldier termites, its camouflage seems to prevent them from recognizing it as an intruder, and they return to the nest. This chemical and tactile camouflage allows the assassin bug to reach into the opening and capture a termite worker, which it kills and consumes by sucking out all the semifluid internal organs, leaving only the exoskeleton.

Such camouflage-assisted prey capture is remarkable enough, but the next step is even more thought-provoking. The assassin bug pushes the empty exoskeleton of its victim into the nest opening and jiggles it gently. Another termite worker seizes the corpse as part of a normal behavior pattern of devouring the body of a dead sibling or carrying the corpse away for disposal. The assassin bug pulls the exoskeleton of the first victim out with the second worker attached. This one is eaten and its empty exoskeleton used in another "fishing" effort. In one case an assassin bug was observed to thus devour thirty-one termites before moving away with a fully distended abdomen.

When chimpanzees fashion sticks to probe for termites, as discussed below, their behavior is considered one of the most convincing cases of intentional behavior yet described for nonhuman animals. When McMahan discovers assassin bugs carrying out an almost equally elaborate feeding behavior, must we assume that the insect is only a genetically programmed robot incapable of understanding what it does? Perhaps we should be ready to infer conscious thinking whenever any animal shows such ingenious behavior, regardless of its taxonomic group and our preconceived notions about limitations of animal consciousness.

Tool Use by Mammals

Among mammals, several scattered examples of simple tool use are known. Polar bears, both in captivity and in the wild, throw sizable objects, and they have occasionally been reported to throw pieces of ice at resting seals, killing or injuring them or at least facilitating their capture (Beck, 1980). But by far the most impressive example, outside of the primates, is the use of small stones by sea otters to detach and open shellfish (Kenyon, 1969). These intelligent aquatic carnivores feed mostly on sea urchins and mollusks. The sea otter must dive to the bottom and pry the mollusk loose with claws or teeth, but some shells, especially abalones, are tightly attached to the rocks and have shells that are too tough to be loosened in this fashion. The otter will search for a suitable stone, which it carries while diving, then uses to hammer the shellfish loose, holding its breath all the while. The otter usually eats while floating on its back. If it cannot get at the fleshy animal inside the shell, it will hold the shell against its chest with one paw and pound it with the stone. The otter often tucks a good stone under an armpit as it swims or dives. Although they do not alter the shapes of the stones, they do select ones of suitable size and weight and often keep them for considerable periods. The otters use tools only in areas where sufficient food cannot be obtained by other methods. In some areas only the young and very old sea otters use stones; vigorous adults can dislodge the shellfish with their unaided claws or teeth. Thus it is far from a simple stereotyped behavior pattern, but one that is used only when it is helpful. Sea otters sometimes use floating beer bottles to hammer open shells. Since the bottles float, they need not be stored under the otter's armpit.

One of the most outstanding examples of tool use, mentioned in Chapter 1, is the chimpanzees' use of probes to gather termites from their mounds. The chimpanzee prepares a probe by selecting a suitable branch, pulling off its leaves and side branches, breaking the stick to the right length, carrying it—often for several minutes—to a termite mound and then probing into the openings used by the termites. If the hole yields nothing, the chimpanzee moves to another one. Even after the tool has been prepared, its use is far from stereotyped. When curious scientists try to imitate the chimpanzees' techniques, they find it rather difficult and seldom gather as many termites. It is especially interesting that the young chimpanzees seem to learn this use of tools by watching their mothers or other members of their social group. Youngsters have been observed making crude and relatively ineffectual attempts to prepare and use their own termite probes. Unlike some of the cases of insect behavior discussed earlier, the termite "fishing" of chimpanzees gives every evidence of being learned.

Many of these tool-using animals must search out, pick up, and sometimes modify a naturally occurring object and then carry it to where it can be used to obtain food. Beck (1980, 1982) argues that tool preparation and use provides no more convincing evidence of intentional thinking than other behaviors, such as shell dropping by birds, that do not qualify as tool use according to his definition. Whether one uses the term *tool* or calls this specialized food-getting behavior by another name, it involves actions quite different from directly seizing, battering apart, and eating the food. The animal carries out a series of actions that are related to eating only indirectly. A gull that picks up a whelk and flies with it to a rocky area before dropping it and flying down, either to pick up the edible fragments or to repeat the process, has certainly separated the preliminary stages of food acquisition from the actual consumption.

One can argue that storing food or carrying it from where it is gathered to a safe place also involves deferring the actual consumption. In all these cases, to varying degrees, the animal goes through a specialized sequence of actions to obtain food, but at least in the early stages, the actions do not involve eating. The differentiation of behavior into these distinct but related stages suggests, though it certainly does not prove, that the animals intend to gather food and realize that grasping a wet and knobby whelk and flying several hundred meters with it will lead to the satisfaction of hunger. Many other sorts of tool-using behavior similarly provide a number of thought-provoking and pertinent examples that suggest conscious thinking.

When comparing chimpanzees with weaver ants or assassin bugs, we must remember the enormous difference in the size and versatile complexity of their brains and behavior. Nevertheless, at the level of neurons and synapses, the fundamental units of all central nervous systems, there are only very minor differences between insects and anthropoid apes or human beings. Unless our understanding of neurophysiology undergoes a major revolution through the discovery of specialized "consciousness neurons" or biochemical substances uniquely linked to conscious thinking, we must assume that the essential difference between central nervous systems that do and do not enable conscious thought are at the level of interactive organization rather than at the cellular level. Whatever the organizational properties of interacting neurons that lead to conscious thinking, they could occur in the central nervous systems of almost any multicellular animal.

To be sure, the central nervous system of a sea otter or chimpanzee is thousands of times larger than that of any insect, but it is arbitrary to assume that an absolute distinction can be made. The differences are more likely to be of degree rather than kind. The use of tools by chimpanzees and sea otters is not enormously more complex than the construction and repair of cases by caddis fly larva or the camouflage and baiting behavior of assassin bugs.

But these scattered examples, selected primarily because more is known about them than about the myriad of other activities carried out by the same animals, constitute only the tiniest fraction of the full repertoire of behavior in adult social insects, birds, and mammals. If a caddis fly larva thinks about its case and what properties make such a case feel right, this is probably one of a few things such a simple creature is aware of. Chimpanzees and sea otters, on the other hand, carry out many other complex activities.

Bowerbirds

A group of closely related birds in Australia and New Guinea construct elaborate bowers of leaves, moss, and branches, which they then decorate by placing conspicuous, often brightly colored, objects in or near the structure. Although the materials used in construction are somewhat similar to those employed in nest building, the function of this behavior is entirely different. These bowers are built by male birds in breeding condition, and they are clearly part of the elaborate courtship behavior. The bowers of many species are covered structures, but others are piles of vegetation on one or both sides of a horizontal twig or branch where the male spends much of his time. Male bowerbirds do not have conspicuous plumage, indeed they are quite drab and unimpressive. By constructing and decorating bowers to attract females, they seem to be making up for their lack of splendid plumage. The males spend most of their time during the mating season constructing, maintaining, and decorating their bowers, which are clearly part of a competitive display between males. Although the bowers are located at some distance from each other, males may steal materials from other bowers to use in their own.

One of the most impressive aspects of the bowers is how they are cleared of other vegetation and decorated. The male bird arranges bright leaves, shells, pebbles, feathers, or other objects in piles or spreads them out in a conspicuous display. Sometimes flowers or colored fruits are included, and in some species the bird colors the walls of the bower by rubbing them with a piece of fruit pulp or some other colored material. Often conspicuous human artifacts are selected, such as bits of bright plastic, tin cups, or auto keys.

Female bowerbirds are small and inconspicuous and therefore

difficult to observe in the thickly forested areas where most of them live. But it is known that the females visit bowers, probably inspecting several before choosing one, where mating takes place. After mating, the female leaves the area and lays her eggs in a rather inconspicuous nest, which she builds in a wholly different location. Male bowerbirds seem to take no part in the care of the young.

Bower building and decorating have been described in considerable detail in the fascinating book *Bower Birds,* by the late H. A. Marshall (1954), a distinguished Australian biologist. He pointed out that bower building is a part of the birds' repertoire of reproductive behavior stimulated at least in part by male hormones and argued that it is not an artistic or aesthetic endeavor. But toward the end of his book Marshall admits that the birds may well enjoy the bowers they build; he emphasizes their strenuous efforts to clean them and to keep the decorations conspicuous and conforming to a certain pattern, for instance, by replacing flowers after they wilt. This pattern varies from male to male, although it is generally similar across a species. As von Frisch points out in *Animal Architecture,* it would be difficult to deny that impressing females is a motive in some human artistic creations. Therefore we should not be too certain that when building and decorating his bower a male bowerbird does not think about what he is doing, the males with which he is competing, and the females he hopes to attract.

Beaver Engineering

Beavers are large aquatic rodents and, as everyone knows, they cut large trees and construct impressive lodges and dams, which usually create sizable beaver ponds. Rue (1964) has written a balanced general account of the natural history of beavers, and their behavior has been thoughtfully analyzed by Wilsson (1968, 1971). Although they range as far south as the southern United States, beavers characteristically live in cold climates where winter conditions are harsh. One of the main functions of a beaver dam is to create deep water so that the beavers can move about in winter by swimming under the ice. Their teeth, with which they cut down sizable trees, are potent weapons, but on land beavers are vulnerable to predation by wolves and other large carnivores. They normally do not stray far from fresh water, and the danger of predation is probably the ultimate, evolutionary reason. Heavy trapping decimated their populations in the nineteenth century, but since then beavers have been protected from overexploitation and have spread into many areas. Wolves and other serious predators have been eliminated as a serious threat, but beavers still refrain from exploiting their favorite food trees if these grow more than two or three hundred meters from the water.

Beavers obtain their nourishment from the bark of trees, together with leaves and tender shoots. Some tree species, such as poplars and aspen have much more nourishing bark and are strongly preferred, but when they are unavailable, beavers eat the bark of many other trees. Their food stores do not seem significantly different in kind from those of other animals, except that the beavers tow medium-sized branches, with their small branches and bark uneaten, to storage piles, many of which are underwater. In winter in cold climates they swim out under the ice from their lodges or burrows to the food stores.

When beavers cut down large trees, they do not seem to eat the thick bark or the chips, so felling a sizable tree, which takes many hours, does not yield food during that time. This labor is ordinarily spread over several nights, and the beaver doubtless eats between tree-cutting sessions. The small branches that will be eaten after the tree falls are far out of reach. Does the energetic animal think about the food it will obtain by its prolonged efforts?

Depending on the local situation, beaver shelters may range from simple burrows in the banks of streams or lakes to the familiar lodges that form artificial islands containing a nesting chamber and underwater entrance. A lodge is sometimes built in a natural body of water, but most often in a pond created by the beavers. The construction of lodges is not unique to beavers; muskrats build simpler and smaller lodges and use them for much the same purpose. As with the construction of any burrow or shelter, the primary function of gathering materials and forming them into a useful structure is to shelter the animal in the future. The nest or burrow is of no use until it is nearly completed, and the process may require many days or weeks of effort. It would seem useful for the animal to understand the end result of its sustained efforts.

Beaver dams differ qualitatively from any other environment-altering behavior known in nonhuman animals. The specific acts of bringing branches, mud, small stones, and sometimes other objects to one part of the stream bed are only distantly related to the future benefit of having a pond deep enough for protection from predators and for travel from lodge to food stores under the ice. While they construct the dam, do the beavers think about the future usefulness of the structure? Scientists firmly believe that dam building is an instinctive behavior pattern and thus cannot be accompanied by conscious anticipation of the results. But we know too little to justify firm conclusions.

Finished beaver dams are quite impressive, often up to a meter high and 30 meters or more long, creating ponds that otherwise would not exist. But a beaver dam is not built in an entirely systematic and efficient fashion, and signs of imperfect planning are often cited as evidence that the beavers do not consciously know what they are doing. One frequently finds miniature and hopelessly ineffective damlets in an area where beavers have been active, sometimes at a place where no pond would be created even if the dam were enlarged, because the surrounding terrain does not rise appreciably above the level of the stream. Or immediately downstream from a large and effective beaver dam a second, smaller dam may be begun that, even if completed, would enlarge the pond only a little. Some of these useless damlets may be the work of young and inexperienced animals, but the subject has not been studied thoroughly enough for us to be sure. Again the information available has probably been severely limited by ethologists' lack of interest in behavior that suggests conscious thinking and planning.

Effective dams are built on small streams that lend themselves to damming. Typically a mated pair begins to build relatively late in the season, and the resulting pond serves for protection and food storage during the winter. The large beaver dams that create sizable ponds often house more than one family and seem to be built by many animals over several years. There is no evidence that beavers start building the largest dams the way a human engineer might, by building up parts of the dam above the rest and finishing the entire structure by closing a small remaining opening. This might not be an effective technique for beavers, since they cannot easily cope with a large volume of rapidly flowing water. The dams seem to be built up gradually by adding material all along their length and extending the dam at the edges when the water rises high enough to flow around it.

Beavers are especially sensitive to the sound of running water, which certainly aids them in repairing their dams. In experimental situations arranged to test the importance of sounds in eliciting dam-repair behavior, beavers piled material on or near loudspeakers playing the sounds of running water or other broadband sounds (Wilsson, 1971). But it seems unlikely that the entire process of dam construction can be explained on the basis of reaction to a single kind of stimulus. In natural conditions they place much material where water is seeping out rather quietly, and a rather noisy stream draining into a beaver pond or running with much turbulence and sound immediately downstream does not stimulate them to build a dam.

As beavers begin to build a dam, do they realize consciously that they will create a pond? Many of the motor actions of dam building appear spontaneously in young beavers that have never seen a dam or old beavers building one, but this is not as crucial a negative argument as is often supposed. A genetically programmed robot beaver might simply create a number of piles in the middle of a stream. Building a dam requires something more, although the innate tendency to gather and pile materials is doubtless an important component of the behavior.

When beavers live in existing lakes and deep streams (including artificial ponds dammed by human engineers), they build no dams. Instead they dig burrows that have an underwater opening to secure and dry nesting chambers underground above the water level. Sometimes they pile branches, mud, and other material on the shore above such a burrow. This variability suggests that something about a shallow stream in an area otherwise suitable for their life style induces dam building. The building stops when the pond is reasonably deep or continues only in a minimal way for maintenance or repair, implying that they recognize when a suitable pond has been created. If an experimental ethologist added more material as a pair of beavers began to build a dam, I suspect they would stop when an appropriate pond had been produced. This question could be put to an empirical test, and if the outcome agreed with my expectation, it would provide further evidence that dam building is more than an automatic and unmodifiable response.

Clearly the "engineering" of beavers presents a challenge to cognitive ethologists, because it makes no sense except as an improvement of the environment that will benefit the animals at a time in the rather distant future. Such behavior would be facilitated if the beavers could visualize the future pond, or if a perceptual template corresponding to the properties of a pond were present in their brains. The dam building itself might be an effort to make the environmental situation match the template, but the beaver would have to be able to anticipate gradual changes over a period of many days or weeks. We have been reluctant to ascribe this ability to a mere rodent, even though beavers are larger than most rodents and clearly capable of more versatile behavior. The comparison between beavers and muskrats is especially interesting, because they would both profit from being able to create ponds, yet the smaller muskrats never make any efforts in this direction.

I do not yet know of any way to ask a beaver whether it contemplates a pond as it drags mud and branches to the middle of a shallow stream. But students of animal behavior have opened up approaches that were totally inconceivable to their predecessors, and some future generation of cognitive ethologists may succeed where we do not yet know how to begin.

Throughout Chapters 3 and 6 we have considered tantalizing information about animal behavior from the viewpoint of the thoughts and feelings that might accompany it. The variety of suggestive evidence is bewildering, and one feels frustrated by the apparent lack of any methods for answering questions I have raised. In the next four chapters I will outline three general approaches, which hold some promise of helping us learn what thoughts and feelings occur in the minds of various animals.

7 Scientific Evidence of Animal Consciousness

Some readers may feel that the preceding chapters have merely posed unanswerable questions and that the behaviorists were right in arguing that scientific investigations can never tell us anything significant about conscious experience in nonhuman animals, if it even exists. The assumption that only learned behavior can be accompanied by conscious thought turns out to be as uncertain as most of the other opinions we have considered in some detail. If learning is not a necessary condition for consciousness, our one slender reed seems to be broken, and the whole subject reduced to an utter shambles. But I believe that such pessimism is far from justified, and the next four chapters describe some scientific developments that offer realistic hope for firm and convincing data.

This chapter will concentrate on two areas: first, the increasing recognition by comparative psychologists that many of their findings are significant evidence of conscious mental experience in animals, and second, new developments in neurophysiology that offer the possibility of identifying electric potentials in living brains that are closely correlated with conscious thinking in human subjects. If similar patterns are found in other species, they might provide objective indices of conscious thinking in animals.

Experimental Analysis of Animal Cognition

The criticism that behavioristic psychologists have neglected animal thoughts has begun to elicit a promising new response. Several psychologists concerned with animal learning and problem solving under controlled laboratory conditions now claim that they and their colleagues have been investigating the working of animal minds all along, even when behaviorism was dominant (Mason, 1976; Roitblat, Bever and Terrace, 1983; Walker, 1983). But in the same breath they are likely to assure us that all animal thinking, and even most human thinking, is quite unconscious. Mind is redefined as information processing; an analysis of how information is acquired, stored, and retrieved, and how it affects overt behavior, is felt therefore to constitute a completely adequate approach to understanding animal minds. But beyond this defensive reaction to the charge that a central aspect of psychology has been neglected, there is a positive and hopeful aspect to these discussions. This is the increasing recognition that when animals learn to perform new and demanding tasks, they may think consciously about the problems they face and the solutions they attempt or achieve. As experimental psychologists become increasingly concerned with mental experiences in animals, their ingenious experimental methods can easily be adapted to the study of animal consciousness, once the taboo against its consideration is laid aside.

This process has been under way for some time, but it has remained hidden behind a smokescreen of behavioristic terminology. Many of the experimenters really do seem quite interested in the possibility of thinking, even conscious thinking, in the animals they study. But they have been inhibited from saying so directly, even to themselves; the result has been what I have called semantic behaviorism (Griffin, 1981). As the behavioristic taboos are relaxed or ignored, the ingenuity that has enabled experimental psychologists to discover so much about learning and information processing can be redirected toward animals' subjective thoughts and feelings. This is a very hopeful prospect, and the chief barriers to its realization lie in the current mindset of many experimenters. But there is a significant, though still largely unrecognized, intellectual underworld of animal experimenters who often suspect that their subjects could scarcely do what they do without some conscious thinking and subjective feeling about their situation and their efforts to solve problems. We can take full advantage of the rich body of data about animal problem solving that has been gathered by psychologists, without being unduly troubled by their stated reasons for studying animals.

Comparative psychologists who have begun to think, talk, and write about animal thoughts tend to use the term *cognition*. That seems more respectable, because it does not imply consciousness as much as the more familiar word *thought*. Many psychologists actively deny that cognition implies consciousness, and some assert dogmatically that animal cognition is always *un*conscious. No cautious agnosticism here, but a flat, outright, unqualified denial, in an area where all scientists agree it is extraordinarily difficult to gather convincing evidence. Nevertheless, hundreds of talented scientists are actively studying how animals solve problems, the extent to which they deal in generalizations and concepts, and whether we can infer from the results of ingeniously contrived experiments that animals have plans and expectations. Even though semantic behaviorism still frowns on terms with mentalistic connotations, the results and interpretations of experiments are continually providing new and stronger evidence that animals sometimes think consciously (Honig and Thompson, 1982). In short, while most of the scientists who study animal cognition deny any concern with animal consciousness, the fervor of these denials seems to be slowly waning.

Rats and other laboratory animals easily learn that a certain light or sound will be followed by an electric shock. They may cringe or show other obvious signs of expecting the unpleasant shock before it is delivered. They also are able to learn that they can prevent the shock by taking some specific action, such as moving to a different part of the cage or pressing a lever. After learning this so-called conditioned avoidance, the experimental animal continues for long periods to perform the avoidance action, even though it no longer receives any shocks (reviewed by Mackintosh, 1974). It seems reasonable to conclude that the animal knows it will be hurt a few seconds after the warning signal if it does not do what will prevent this from happening. It expects a painful shock following the warning light and anticipates that it will not be hurt if it does what it has learned will be effective. Psychologists are reluctant to describe this sort of conditioned avoidance in mentalistic terms, preferring a behavioristic account of what happens.

Many students of animal learning have noted that animals often act as though they are expecting something, and if it does not become available they appear surprised or disappointed. Tolman (1932, 1937) emphasized this sort of behavior in rats that were required to learn complex mazes in order to obtain food. In a typical experiment, after the rat had learned a moderately complex maze and was performing almost perfectly, choosing correctly a long series of right or left turns, the experimenter withheld the reward. On reaching the goal box and finding no reward, the rats would appear confused and search about for the food they had reason to expect.

One of the most dramatic examples is still the one described by O. L. Tinklepaugh (1928), who trained monkeys to watch the experimenter place a favorite item of food, such as a piece of banana, under one of two inverted cups that remained out of reach until a barrier was removed. The purpose of the experiment was to measure how long the monkey could remember which cup hid the piece of banana. When a monkey had learned to do this almost perfectly, in some experiments the banana was replaced by lettuce during the waiting period when the monkey could not see the cups or the experimenter. As Tinklepaugh described the results, the moderately hungry monkey now "rushes to the proper container and picks it up. She extends her hand to seize the food. But her hand drops to the floor without touching it. She looks at the lettuce but (unless very hungry) does not touch it. She looks around the cup . . . stands up and looks under and around her. She picks up the cup and examines it thoroughly inside and out. She has on occasion turned toward the observers present in the room and shrieked at them in apparent anger" (p. 224, also quoted in Tolman, 1932, p. 75).

Numerous other experiments have confirmed Tolman's thesis that animals act as though they expect a particular outcome at certain times. This trend has been well summarized by Walker (1983): "Some kind of mental activity is being attributed to the animals: that is, there is considered to be some internal sifting and selection of information rather than simply the release of responses by a certain set of environmental conditions. Knowledge of goals, knowledge of space, and knowledge of actions that may lead to goals seem to be independent, but can be fitted together by animals when the need arises" (p. 81). Naturalists and ethologists have gathered abundant evidence that such needs do arise very commonly in the natural lives of animals, and the resulting behavior strongly suggests that they understand in an elementary fashion what the problems are and how their behavior is likely to solve them. Animals appear to think in "if, then" terms. "If I dig here, I will find food," or "If I dive into my burrow, that creature won't hurt me." Likewise in the laboratory, "If I peck at that bright spot, I can get grain," or "If I press the lever, the floor won't hurt my feet."

A relatively simple case is the ability of numerous animals, including many invertebrates, to learn that food is available in a certain place at a certain time of day. They will return to this place at or shortly before the appropriate hour on subsequent days and may continue, though with decreasing regularity, even after many days when no food has been found there. Such experiments are ordinarily interpreted as evidence of animal learning and memory and of the presence of biological clocks. Obviously something in the animal causes it to act on the basis of information received sometime in the past. Now that we have, in the words of the philosopher Daniel Dennett (1983) "cast off the straitjacket of behaviorism and kicked off its weighted overshoes," it seems reasonable that these animals really do expect food at a certain time or place and that they experience disappointment, annoyance, or other subjective emotions when their expectations are not fulfilled.

There is abundant evidence that many animals react not to stereotyped patterns of stimulation but to *objects* that they recognize, despite wide variations in the detailed sensations presented to the animal's central nervous system. As reviewed in Chapter 4, a Thompson's gazelle recognizes a lion when it sees one. The lion's image may subtend a large or small visual angle on the retina, and it may fall anywhere within a wide visual field; the gazelle may see only a part of the lion's body from any angle of view. Yet to an alert tommy, a lion is a lion whether seen from the side or head on, whether distant or close, standing still or walking. Furthermore, its perceptions of lions are obviously separated into at least two categories: dangerous lions that act as though ready to attack, and others judged to be less dangerous on the basis of subtle cues not obvious to a human observer without considerable experience. Comparable behavior is so common and widespread among animals living under natural conditions that it seems not to require any special scientific analysis. Yet the ability to abstract the salient features of a complex pattern of stimulation, often involving more than one sense, requires a refined ability to sort and evaluate sensory information so that only particular combinations lead to the appropriate response.

Such discriminations have been extensively studied in laboratory animals, many of which are adept at recognizing objects that
they have learned are important, regardless of size or distance,
and even if only a part of the object is visible. But in laboratory
studies of animal learning and discrimination, classification of
sensations is often studied by a sort of mental inversion of the
approach outlined above (reviewed by Mackintosh, 1974). Instead
of asking how the animal recognizes an object from a wide variety
of viewpoints, and even with some distinguishing features partly
removed, experimenters first train animals to respond to a single,
simple stimulus and then present them with other stimuli that
differ from the first along well-defined dimensions, such as wave-
length, sound frequency, intensity, or size. It is scarcely surprising
that animals trained to seek food at the sound of a 1200 Hz tone
also go to the food dish on hearing 1150 Hz, even though they
can easily detect this frequency difference. The results of such
experiments are interpreted as demonstrating "gradients of generalization"; the rate or intensity of response diminishes as the
stimulus is increasingly different from the one the animal originally learned. But while such experiments have a satisfying simplicity and tidiness, they do not begin to explain how animals
discriminate between two or more categories of real objects that
may vary widely from one presentation to another and that must
be distinguished by relatively minor cues that tell, for instance,
whether the object is edible.

An especially significant sort of learning occurs when one animal imitates the behavior of a companion after watching it obtain
food. This observational learning has frequently been seen in
captive apes and porpoises, as described in more detail in Chapter
10. It also occurs in other mammals and in birds, such as the tits
studied by Krebs and his colleagues, as discussed in Chapter 3.
Additional examples of observational learning in birds are described by Alcock (1969), Curio, Ernst, and Vieth (1978a, 1978b),
Mason and Reidinger (1982), and Huang, Koski, and de Quardo
(1983).

One of the best-studied examples is the vocal learning of songbirds. Although observational learning suggests visual scrutiny,
auditory observation and imitation are equally important. If an
experimenter plays tape recordings to young male birds, when these
birds reach sexual maturity and begin to sing, they will copy the
recorded songs. Females of some species respond more strongly

to male songs that resemble those they heard when they were quite young (Marler, 1970; Marler and Peters, 1981; Kroodsma and Miller, 1983). Learning to copy the details of another animal's behavior, including its vocalizations, may be aided by conscious thinking about the task.

Humpback whales are also proficient at vocal learning (Payne, 1983). These whales sing long and complex songs, and whole populations share many distinctive themes, which change gradually from month to month and even from year to year. To keep their complex songs so much like those of the other whales, they must listen to each other and perhaps think about the process of imitating what they hear.

When an animal imitates the behavior of another, it is often difficult to know whether the imitator has simply been stimulated to do something it has done before, a process called social stimulation (reviewed by Clayton, 1978). To be sure that apparently imitative behavior has actually resulted from observational learning, it is necessary to ascertain that the new behavior is clearly different from anything the animal has done before or to control very carefully the previous opportunities for the animal to behave in the way it is supposed to learn by observation. Psychologists tend to prefer experiments in which one animal is studied in isolation, for the good reason that even a single animal presents many difficult problems to be controlled and analyzed, while two interacting creatures multiply the variables considerably. But some of the situations where conscious thinking is most likely to occur are social interactions, predator-prey relations, and other occasions when animals must interact. To analyze observational learning in detail, it might be useful to arrange for the animal under study to observe a representation of another animal, such as a sound movie or video recording. For this to be effective, the animal must react to the reproduction more or less as it would to a real live companion. This approach can be quite difficult to arrange, but it might allow detailed analysis of just what the imitator is observing and learning.

In another type of learning experiment, a rat is trained to recognize that one pattern, such as a triangle, marks the location of food, but that another, say an equally conspicuous circle, does not. Then, after the rat has solved the problem and has been performing quite accurately for some time, the experimenter suddenly changes the rules of the game, so that the circle shows where food is available and the triangle yields nothing. In time the rat learns the reversed rules and again performs almost perfectly; it has changed its searching image from triangle to circle in what is called reversal learning. But an interesting difference results from overtraining some rats on the first problem by letting them make the correct choice dozens or hundreds of times, while giving others just enough training that they are barely proficient. One might suppose that the overtrained rats would have the "triangle marks food" rule so thoroughly drilled into their brains that it would be harder to learn the reversed problem. But careful experiments have shown that under some conditions overtrained rats learn more easily than the others to reverse their choice (Mackintosh, 1974). Perhaps during the dozens of trials after the problem is learned, they begin to think consciously about the two patterns and thus find it easier to grasp the new relationship. No one can say for sure, but communication with the rat via such reversal experiments might be telling us something important.

Mackintosh (1974) and Walker (1983) have reviewed several other types of experiments showing that laboratory animals can learn relatively abstract rules, such as oddity or the difference between a regular and an irregular pattern. In oddity experiments the animal is presented with a number of stimuli or objects, one of which differs from the others in some way; it must learn to distinguish this "oddball" from the other members of the set. For many animals, learning a single case of this sort is not difficult. Chimpanzees, however, have learned to generalize oddity as such, and having learned to select a red disk placed with two blue disks, and a blue disk accompanied by two reds, they also selected the oddball when it is a triangle with two squares. Pigeons have much greater difficulty with comparable problems, but do better than cats and raccoons. Variations on this experimental theme have led to unexpected results. For instance, Zentall and colleagues (1980) compared the performance of pigeons faced with two types of oddity problem. In one case the birds saw a five-by-five array of twenty-five disks, one of which differed in color from the remaining twenty-four. In the other problem there were three disks in a row, two alike and the third a different color. In that case, if the position of the odd-colored disk was varied randomly, or if the actual colors were changed, for instance from one green and two reds to two greens and one red, the pigeons failed to solve the problem. But with the array of twenty-five they quickly learned to peck at the one that differed in color, even when the colors shifted randomly from twenty-four reds plus one green to twenty-four greens and one red.

In a related type of experiment Delius and Habers (1978) trained pigeons to distinguish pairs of visual patterns according to their relative symmetry or asymmetry. Having learned this task, they were also able to make the correct distinction on the first try when given new pairs of shapes, some of which were symmetrical and the others not. Furthermore, Bowman and Sutherland (1970) trained goldfish to distinguish between a perfect square and one with a bump in the top edge. In one of many variations, goldfish that were trained to swim toward a square having a small triangular extension from its top, rather than to a perfect square, also selected a circle with a small semicircular indentation in the upper edge in preference to a plain circle. They seemed to have learned to distinguish simple shapes from the same shape complicated by an indentation or outward bulge. Walker (1983) expresses surprise that "even a vertebrate as small and psychologically insignificant as a goldfish appears to subject visual information to such varied levels of analysis." Why should a comparative psychologist be so surprised? Apparently he is unduly influenced by the faith that only primates, or only birds and mammals, have the capacity for learning moderately complex discriminations. But the natural life of almost any mobile animal requires it to discriminate among a wide variety of objects and to decide that some are edible, others dangerous, and so forth.

Many laboratory studies of learning and discrimination between stimuli employ the Skinner box, in which a very hungry animal, typically a rat or pigeon, is isolated from virtually all stimuli except those under study. To obtain food the animal must manipulate something in the box when a particular stimulus occurs. The devices to be manipulated have been chosen to elicit behavior which the animal normally prefers; rats place their forepaws on objects just above the floor, and pigeons peck at conspicuous spots at eye level on the wall. This then activates a mechanism that makes food accessible. Straub and Terrace (1981) trained pigeons in a Skinner box to peck at colored windows in the wall and to follow a particular sequence of colors. To get its food, one pigeon might have to peck first red, then blue, yellow, and green, while another was required to peck in the sequence yellow, red, green, blue. In the Skinner box the pigeon was faced with two rows of three spots that could be illuminated with different colors. In the most significant experiments four of the six spots were illuminated simultaneously, each one a different color, but the positions of the colors varied from trial to trial; the pigeon had to ignore the position of the spots and select the appropriate color in the correct sequence to obtain food. Several pigeons learned to perform this task at a level far above chance, indicating that they had learned a sequential rule that guided their decisions about which spot to peck. It seems possible that they thought something like "I must peck at red first, then blue, next yellow, and then green."

Even more impressive and suggestive experiments by Herrnstein, Loveland, and Cable (1976) and Herrnstein (1979, 1982), have demonstrated that pigeons can learn the somewhat abstract concepts of discriminating between pictures that do or do not contain a general category of object, rather than a specific pattern. Hungry pigeons learn to peck at colored pictures projected on a miniature screen in one wall of their Skinner box. Pecking at the correct picture yields food, and other pictures yield nothing. Pigeons can remember for years that a particular picture, such as an aerial view of a certain area, is rewarded in contrast to other scenes. In recent experiments the pigeons faced a much more difficult task, because each picture was shown only once. Dozens or hundreds of color slides were projected, and food was provided only when the pigeon pecked at those that included a particular feature. In a simple problem of this sort, the rewarded pictures might all include some bright red object. Even though the actual scenes differed enormously, it is not surprising that the pigeons could learn to peck at those containing red colors. In the most significant experiments the criteria were much more complex. All the positive, rewarded, slides would show trees or parts of trees, while the negative pictures would include lampposts, vines growing over walls, and other treelike patterns, but no whole trees, branches, leaves, or tree trunks. Pigeons learned to choose the positive examples. In other experiments of this type, pigeons have learned to distinguish pictures showing a particular person dressed in various types of clothing, although her face was always visible. Many of the negative examples showed other people, some even dressed similarly to the person whose presence signaled "peck at this one to get food." From such experiments Herrnstein, Loveland, and Cable (1976) concluded that pigeons can learn to recognize some natural categories even when the size or angle of the representation changes or when it is mixed with a bewildering variety of other objects.

These discriminations are not easy for the pigeons; many months of training are necessary, and the final performance, while clearly much better than chance, is not perfect. But the bird must have developed some concept of what to look for in deciding which pictures warrant a hopeful peck. Can they do this without any conscious thinking? If the criterion is trees (Cerella, 1979, 1982), or pigeons versus other birds (Poole and Lander, 1971), this seems to be something a pigeon might naturally think about and look for, but natural selection has not prepared pigeons for the searching image of a particular woman. In another experiment Herrnstein and de Villers (1980) used underwater photographs, some of which contained a fish while others did not. Pigeons learned this discrimination too, although neither they nor their ancestors for millions of years had had any experience with underwater scenes. Thus these results must be based on a general ability to learn what features are common to one set of highly varied and complex pictures and lacking in another set. A person faced with this task would consciously think about the common features present in those he knew were correct choices, and when he solved the riddle, he would probably be conscious of the insight "Ah, it's the pictures with trees that I'm supposed to pick out."

Within the next few years the enormous body of data resulting from experiments on animal learning, problem solving, and cognition may well be reconsidered in a new light, by thoughtful psychologists interested in conscious thinking in animals. The initial stages of this development have been reviewed by Walker (1983). Along with this, we can hope for experiments designed to measure as objectively as practicable the extent and the content of this conscious thinking. Boring (1950) pointed out that experiments on learning and problem solving are in effect a type of communication between the psychologist and the experimental animal. The experimenter asks questions by arranging the situation, and the animal answers through its behavior. But this metaphor is appropriate for our purposes only if the results of the experiment tell us something about the animal's subjective experience as well as its behavioral capabilities. The two often coexist, but they are not one and the same. Furthermore, as explored in Chapters 8 and 10, under the right conditions other, natural sorts of communication provide more direct access to the thoughts and feelings of animals.

Electrical Indices of Thinking

The principal way in which neurophysiologists monitor the activity of a neuron or even a whole brain is by recording the electrical potentials that accompany both the transmission of nerve impulses among neurons and the processes in which some impulses cross from one neuron to another at synaptic junctions. To monitor the activities of single neurons, microelectrodes are thrust into brain tissue, and these pick up characteristic signals that accompany—and are perhaps almost the same as—each nerve impulse. In the many detailed studies on monkeys of the action potentials from single neurons, neurophysiologists have sometimes monitored a neuron that is activated most strongly by a moderately complex stimulus, such as an outline of a monkey's paw (Gross, Rocha-Miranda, and Bender, 1972; Gross, Bender, and Rocha-Miranda, 1974; Gross and Mishkin, 1977). Unfortunately these experiments have not yet been carried nearly far enough to offer much hope that they can develop into a method of tapping an animal's thoughts directly.

Another type of electrical potential in human and animal brains can be recorded from the outer surface of the scalp. Technically these weak electrical signals are known as electroencephalograph (EEG) signals or, more popularly, as brain waves. They have a very much lower voltage than the electrical signals recorded from electrodes inside the brain. But all neurophysiologists agree that the EEG waves are similar in basic properties and origins to some of the low-frequency potentials that can be recorded within the brain. EEG potentials are obviously studied much more easily and with no possibility of injury or even discomfort to the subject.

EEG signals in mammals and other vertebrates can be recorded in much the same manner, although in cats and monkeys the skull, superficial muscles, and skin present a more troublesome electrical barrier than in human subjects. Electrodes placed just under the skull of an experimental animal (the surgery and placement being done under anesthesia) pick up essentially the same EEG potentials as those recorded from the human scalp. Practical difficulties have limited comparative studies of birds, reptiles, fishes, or invertebrate animals, principally because the magnitude of the potentials from their central nervous system is relatively small compared to competing electrical potentials from muscles or other sources outside the brain (Klemm, 1969). But when these practical difficulties can be surmounted, similar electrical signals can be recorded from the central nervous systems of all vertebrates.

In EEG recordings the strongest electrical signals are recorded when the brain is in a relatively inactive state, when human subjects have their eyes closed and are not moving actively or thinking about anything in particular. In some ways this limits the usefulness of EEGs in the study of the more complex and, for our purposes, significant activities of human or animal brains. But by refined methods of analysis EEG waves recorded from awake human subjects can show differences in activity in the right and left cerebral cortices as the subject thinks about spatial relations or about verbal problems such as selecting synonyms from a list of words (Ornstein et al., 1980).

Even in sleep there are prominent EEG signals, but they are lower in frequency than the most prominent so-called alpha rhythm, with a frequency of about 5 to 8 Hz or waves per second. EEG potentials are believed to result from the nearly synchronous activity of many thousands of neurons and synapses. In a more active brain this synchrony is reduced or absent. When a person is lying quietly with his eyes closed and not engaged in any vigorous mental activity, the alpha waves are likely to be quite prominent. If the person is then asked to do some mental task, such as solving arithmetic problems, the waves diminish in amplitude and become more difficult to distinguish from other, irregular waves of undetermined origin. This observation, made in the 1930s, suggests that whatever the alpha rhythm may reflect, it is not a helpful measure of conscious thinking.

Neurophysiologists later discovered that discrete sensory stimulation, such as a flash of light or a brief sound, also elicits small electrical potentials that can be recorded at the human scalp. Ordinarily these are too small to be detected reliably against the background of other electrical signals, but this difficulty has been surmounted by repeating discrete stimuli over and over while the electrical voltages recorded at the scalp are averaged, usually by a computer. The resulting averaged evoked potential is best understood by means of a graph that displays voltage at the scalp as a function of time elapsed since the presentation of the stimulus. Often it is necessary to average hundreds or even a few thousand stimuli, which ordinarily are presented at a rate of perhaps one per second. Thus the practical limits of how long a subject is willing to participate in such an experiment restrict the number of presentations and the effectiveness of the averaging process.

In spite of these technical problems, it is possible to accurately and consistently record complex waves describing the fluctuations in electrical potential that follow a brief and discrete sensory stimulation. These need not be limited to light or sound; significant experiments have been conducted with light tactile sensations or feeble and painless electric shocks to the skin. But a stimulus that takes several seconds to deliver is not useful because the averaging process requires a discrete temporal starting point. Also the stimulus must be rather brief; otherwise, excitation from the latter part might complicate the electrical signals generated by its initial portions. The significance of evoked potentials has recently been analyzed by Callaway, Tueting, and Koslow (1978), Desmedt (1981), Galambos and Hillyard (1981), Rockstroh and colleagues (1982), and Hillyard and Kutas (1983).

The variety of stimuli that can be presented effectively includes flashes of light of different brightness or color, tonal sounds of varying frequency, brief bursts of noise, or clicks. The sounds can be spoken words, and the lights can be pictures, including words or letters flashed on a screen for a short interval. The resulting waveforms contain many successive changes in a positive or negative direction, and certain waves are usually much larger than others. But so long as the waves are clearly above the background noise, their amplitude is usually less significant than their timing and their relationship to what the subject's brain is doing.

The time intervals after the brief stimulus are usually measured in milliseconds. One of the more widely used terminologies identifies peaks as electrically negative or positive, and gives the time in milliseconds after the onset of the stimulus. The first waves after a stimulus, such as a light or sound, which occur within about 1 to 50 milliseconds, seem to reflect the activation of the more peripheral levels in the visual or auditory system. They do not vary greatly according to the subject's mental or neural activity and are not greatly diminished during light anesthesia. These early potentials quite clearly reflect sensory input to the brain, for very similar potentials can be recorded with electrodes deep inside the brain tissue at locations known to be concerned with early stages in the processing of sensory excitation. For this reason the early potentials are of much less interest to us than those that occur 100 to 500 milliseconds after the stimulation.

The potentials having relatively longer latencies are often called "event related potentials," or ERPs. Although this term is some-times applied to short-latency evoked potentials, I will follow the usage of designating as ERPs potentials that seem to be associated with more complex brain activities. In human and other mam-malian brains most of the more complex processing of sensory information, and its use for purposes that approximate thinking, occurs in the cerebral cortex, which is why electrodes to record EEG waves or ERPs are placed on the top of the head.

Human ERPs are complex and have been studied in a wide variety of experiments, but I will concentrate on a few of the more prominent and better-studied waves. One of the most in-teresting is known as the P300 wave, meaning that it is positive and peaks at approximately 300 milliseconds after the stimulus onset. This wave seems to reflect complex information processing within the brain, and possibly something like conscious thinking, although that is not clearly established. One reason for linking it to conscious thought is that although the P300 wave ordinarily results from a discrete sensory stimulus, it has been shown to occur when the subject is expecting a stimulus that does not occur. The P300 wave is little affected by the intensity of the physical stimulus, provided it is intense enough to be clearly detected at all. It is also significant that the P300 wave varies in amplitude according to what the subject is doing with his brain. I use the phrase "doing with his brain" to maintain a neutral position as to whether these brain activities are accompanied by conscious thinking, but pertinent evidence on this point will be discussed below.

One of the earliest experiments indicating a possible relation-ship between the P300 wave and thinking was conducted by Sutton and colleagues (1965). In more recent and refined experiments by Picton and Hillyard (1974), subjects were presented with a long series of uniform clicks, one of which was occasionally omit-ted. The P300 wave that occurred about 300 milliseconds after an expected click was omitted often had a larger amplitude than after other clicks in the series, indicating that the immediately preceding stimulus is not necessary for the brain activity leading to the P300 wave.

In another experiment subjects were presented with light flashes and then, in half of the presentations, at random, with rather faint tones, close to the threshold level at which the subject could no longer detect them. The subject was asked to report whether he heard the tone or not, and the test was difficult enough that he frequently made mistakes, failing to report a tone that actually did reach his ears or reporting a "false alarm" when no tone was in fact delivered. There were thus four possible results: correct identification of a tone, correct judgment that none had occurred, a failure to detect an actually present stimulus, and a false alarm. Of these four categories, only the correct judgment that the tone actually occurred yielded a prominent P300 potential. In recent variation on this experimental theme, subjects were given a series of stimuli including two signals, one occurring much less frequently than the other. This relatively rare one is called the oddball stimulus. The subject was required to make some simple response, such as pressing a key, when the oddball stimulus was noticed. Under the conditions of these experiments all stimuli produced P300 waves, but the oddball stimulus usually produced a larger one (Galambos and Hillyard, 1981).

Even more interesting were experiments in which the stimuli were words that the subject had to think about in order to give a correct response (Donchin, 1981). In one experiment the name *David* was presented 80 percent of the time and *Nancy* in the remaining 20 percent. In another experiment a number of different but familiar first names were used, 80 percent being masculine names and 20 percent feminine. In still another variant, 20 percent of the words rhymed with *cake* and 80 percent did not. In a final arrangement the oddball words were synonyms of *prod;* the others were not. The subject was asked simply to count the number of times he perceived the rare stimuli. In all cases there was a prominent P300 potential, but the rare stimuli elicited, on the average, larger P300 potentials with a longer latency than the other stimuli. The latency difference between the P300 waves in response to frequent and rare stimuli was greater when the problem was to decide whether the stimulus was synonymous with *prod,* rhymed with *cake,* or was a feminine rather than a masculine first name, than when the subject had only to distinguish between the two words *David* and *Nancy.* The brain requires a little more time to make these more complex discriminations.

This sort of experiment has been varied in a number of ways, but these examples are sufficient to show that the P300 wave is strongly correlated with what can be conservatively called information processing by the human brain (Chapman et al., 1978; Kutas and Hillyard, 1980; Duncan-Johnson and Donchin, 1982). Do experiments of this sort provide an objective, measurable electrical index of whether a person is thinking about a simple discrimination problem? The cautious scientists who have carried out these experiments are reluctant to jump to such a heady conclusion. Galambos and Hillyard (1981) selected the conservative title "Electrophysiological Approaches to Human Cognitive Processing." Yet the experimental results certainly suggest that ERPs provide a crude index of conscious thinking. Why are the scientists directly concerned with these experiments so cautious? Their reasons are many, but the following are probably the most important. First of all, scientists have learned to be skeptical when something that seems so exciting appears on the horizon. In the late 1920s, when amplifiers first became sensitive enough to record EEG potentials, many extravagant claims were made that brain waves would allow electrical measurement of all sorts of brain functions. The more recent discoveries that ERPs correlate in a simple way with discriminations about the nature of stimuli may be seen as opening a door to electrical monitoring of human thought, but such a revolutionary possibility is rightly treated with great caution by the careful scientists who do these experiments.

Another difficulty is that the relatively slow electrical changes that can be recorded from the human scalp provide only a very blurred picture of the activities of neurons within the human brain. For a potential to be recorded from the scalp, thousands or even millions of neurons and synapses must be active in approximate synchrony. Otherwise their electrical signals, recorded at a distance many times the size of the cellular structures from which they arise, offset each other, so that even when millions of brain cells are active asynchronously, virtually nothing may be measurable at the scalp. Many neurophysiologists (for instance, Mountcastle, 1981) feel that scalp recordings provide only a very coarse measure of brain function, even though they do show significant and reproducible correlations with some types of thinking. Neurophysiologists tend to be more satisfied when the discrete electrical signals from individual cells can be recorded with microelectrodes thrust deep within the brain tissue. Obviously this cannot be done with human subjects except under very exceptional circumstances during brain surgery, in order to diagnose the nature of brain disease or damage.

Since probing of human brains with exploratory microelectrodes for purposes of studying neurophysiology is out of the question, we need indirect evidence to determine what portions of the brain give rise to the ERPs with such interesting properties. This effort has not yet progressed very far, probably because quite large areas of the cerebral cortex, and perhaps subcortical areas as well, are involved. The brain processes activated in such discrimination problems may be widely diffused throughout the cortex and hence it may not be possible to localize them to the nearest millimeter.

By now readers are probably wondering impatiently whether similar experiments have been carried out with nonhuman animals, and if so, what the results have been. Does the brain of a monkey, a cat, or another animal give out anything comparable to the human P300 wave? Animal experiments have only recently begun to answer this important question. Although no experiments directly comparable to those described above have yet been carried out, it has been found that some simple discriminations by animals do result in ERPs that resemble in many ways the human P300 (reviewed by Galambos and Hillyard 1981).

Electrodes in the brain of a cat can record similar long-latency ERPs not only when stimuli are presented repetitively, but when an expectable stimulus is omitted (Wilder, Farley, and Starr, 1980; Buchwald and Squires, 1982). In other experiments the ERP in response to a light flickering 7.7 times per second increased in amplitude and changed in waveform when the cat had been trained to make a simple response in order to avoid a shock that would otherwise follow this stimulus. After learning to avoid the shock in this way, the cat was further trained that it could reach food by approaching a light flashing at 3.1 times per second, whereas it should move away after 7.7 flashes per second to avoid a shock. Again the ERPs increased somewhat in amplitude and changed to a different waveform after the cat had learned to respond differently to the two rates of flashing.

In another experiment, reviewed by Galambos and Hillyard (1981), Neville and Foote (in press) have attempted to make direct comparisons between ERPs in human subjects and monkeys. Some of their stimuli were brief tones, 190 milliseconds in duration. When these were repeated over and over, neither the human subject nor the monkey showed an especially prominent P300 wave. But when one tone in approximately twelve had a distinctly different frequency, the oddball stimuli produced the typical P300 wave in the human subject and, in the monkey, a slow positive wave having a somewhat different waveform, from about 320 to 550 milliseconds after the stimulus. In both subjects the response to the oddball stimulus was distinctly larger than the ERP following the more common signal. In another version of this experiment a third type of stimulus resembling a brief dog bark occurred 8 percent of the time, a tone of a different frequency another 8 percent, and the customary tone in 84 percent of the trials. The dog bark was selected because it differed much more drastically from the common stimulus than a tone that differed only in frequency. Both the human and monkey recordings showed a peak at about 300–350 milliseconds, although the waveforms were different. But these differences could result from the fact that in the monkey the electrodes were actually implanted into the skull and were thus closer to the brain tissue.

These and numerous other experiments indicate that long-latency ERPs occur in cats and monkeys under conditions at least roughly comparable to those recorded in human subjects. But it is premature to conclude that these potentials provide a reliable index of even the simplest type of conscious thinking. One difficulty is that experiments like those of Neville and Foote have not yet been carried out with both human and animal subjects under strictly comparable conditions. Obviously one could not present animals with stimuli that differ according to verbal categories such as men's or women's names. But the stimuli could be within the discriminative capacities of an animal and, more important, truly important in its ordinary life. As cautious scientists overcome their initial reluctance to enter into such potentially sensational lines of investigation, such experiments will almost certainly be carried out.

One facet of human ERPs that needs closer investigation could be studied in discrimination experiments in which the subject sometimes thought consciously about the task and sometimes did not. For instance, after a time a boring and repetitive discrimination between two tones might well come to be performed accurately but without conscious attention. Perhaps the subject could signal when he was thinking about the discrimination and when he was daydreaming or listening to background music. Would the P300 or any other pattern of electrical signals from the brain show a clear difference between the two cases? Published descriptions of earlier experiments suggest that if the subject is not paying conscious attention to the discrimination process, the P300 is greatly reduced or undetectable. But this needs to be tested unequivocally, and if a certain type of ERP proves to be reliably correlated with conscious thinking, neuroscientists can then move on to investigate whether it also occurs in animals.

One limitation of all experiments on ERPs to date has been

that they appear to indicate only the presence or absence of thinking, telling us nothing about the content of the thoughts. An important first step toward understanding a mechanism is learning how to detect when and where it occurs, but even this process has only begun. There is as yet very little evidence that the P300 or any other event-related potential differs according to the content of the discriminative thinking. One step in learning more about content is the demonstration by Chapman, and colleagues (1978) that the waveform of auditory evoked potentials show significant statistical correlations with the emotional properties of spoken words.

The remoteness of scalp electrodes from brain tissue is probably a major reason for the coarseness of the correlation between electrical signals and the corresponding thought processes. Yet the restricted nature of the circumstances in which ERPs can be recorded may also be important. So far ERPs have been measurable only when brief stimuli are repeated hundreds of times to permit averaging and measurement of otherwise undetectable signals. While oddball stimuli produce somewhat larger than average P300 waves or differences in latency, and omitted expectable stimuli also do this under some circumstances, the experimental conditions severely limit the possibility of detecting any differentiation among ERPs. Despite these difficulties, if ERPs do reliably correlate with conscious attention to a sensory discrimination task, this fact could provide a very useful method for examining animal brains for electrical signs of conscious thinking. Neurophysiologists say they are working slowly toward that goal, but that the subject is not yet well enough understood to permit such an ambitiously direct attack.

Many difficulties may complicate the interpretation of the animal experiments I have suggested. Even if animal and human ERPs are shown to differ in latency, waveform, and amplitude, these are of less importance than whether ERPs correlate clearly with the nature of the discrimination problem. It has already been shown that cat and monkey brains give ERPs in response to unexpectedly missing members of a train of stimuli, and in some situations give larger responses to oddball stimuli. But this may simply reflect the fact that such novel or unexpected patterns of sensory input activate larger areas of the animal's brain or more closely synchronized activity, and hence generate larger ERPs. Some experiments show that ERPs from animal brains change when the stimulus becomes important to the animal, as when a cat is trained that the stimulus means some overt motor behavior is required in order to avoid a shock. But it would be even more interesting if the stimulus were more natural, something its brain were prepared by evolution or previous experience to handle efficiently. Suppose, for example, that the signals concerned access to food or a potential mate, or were more closely related to the animal's natural behavior than learning to avoid an electric shock.

Perhaps experimental conditions could be arranged that would be comparable to the human experiments that discriminate between male and female names. It would be especially interesting to use the calls used by highly social animals for social communication, especially since these sometimes also serve for individual identification, at least in monkeys. It is easy for me to outline potentially exciting experiments for my colleagues to carry out, but they will probably prefer to make their own choices. This whole area seems extremely promising, and if the experiments yield consistent and positive results, they might provide, at long last, an objective, verifiable method for detecting the presence of conscious thinking.

8 A Window on Animal Minds

Animals cannot analyze electrical potentials from the brains of their companions, but they do interact with each other in ways that require understanding each other's moods or behavioral dispositions. This understanding is ordinarily achieved by communication. Many types of information received through different sensory channels are involved in animal communication, as reviewed in the volume edited by Sebeok (1977). Because we rely so heavily on vision and hearing, it is easiest for us to understand animal communication involving these two senses. But many species communicate through odors, tactile stimulation, and, in a few specialized cases, quite different sensory channels, such as the highly refined electrical sensitivity of certain fishes (Hopkins, 1974, 1981). For our purposes, the particular communication channel is of secondary importance; we want to know if conscious thinking accompanies the sending and receiving of communication signals between animals.

Many forms of animal communication do not suggest any conscious thinking by the communicator or the recipient. In relatively simple cases, an animal's behavior conveys information to others as an incidental byproduct. When a large and dominant monkey rushes toward a newly discovered item of food, a subordinate companion, who is probably a genetic relative, gets the message that he should stay out of the way or at least refrain from grabbing the food himself. While the subordinate may well think that he should stay away from the food, there seems no need to postulate that the dominant animal plans to communicate some such message as "Get out of my way," since the simple act of moving toward the food suffices to achieve that result. In many other such cases it seems reasonable to assume that one animal, the communicator, gives the recipient animals information simply through its actions. This incidental communication is not of special interest to our search, except that it tells us that the recipients judge what they should do by watching the communicator.

Most animals also carry out specialized communicative behavior which serves primarily, if not exclusively, to convey information to another animal. Sometimes this distinction cannot be made in a casual observation, but in virtually all social animals the degree of specialization of true communicative behavior is quite obvious. An animal approaches or faces another while vocalizing loudly, or making a conspicuous anatomical structure more clearly visible. Often there is a reciprocal exchange of signals through behavior that has no other function. Some scientists have voiced doubts that animal communication is ever more than an accidental byproduct, comparable to groans of pain emitted regardless of the presence of any listener. But such skepticism becomes quite ridiculous when one considers how different this communicative behavior is from anything else the animal does, how explicitly it is directed at other animals, and how the recipients respond. The following examples illustrate this point and show how animal thoughts might be accessible to our scrutiny if we could understand the communication that flows back and forth between animals under natural conditions.

Female Choice and Male Display

Reproduction requires the male and female animal to coordinate their efforts by transmitting and receiving signals of one sort or another. In most species the female must make a much larger biological effort, and the male is ready for mating a larger fraction of the time in most species (though perhaps not our own!). It is therefore helpful for the females to broadcast some sort of signal when she is ready to copulate or produce eggs in coordination with the male's production of sperm. In many aquatic animals fertilization is external, but eggs and sperm must be released into the water in approximate synchrony and in close juxtaposition. But regardless of the reproductive physiology, efficient communication is necessary to bring the two sexual partners together at the right time and place when both are ready.

In many species the female attracts males by releasing a specific scent that is effective at very low concentrations, so that males detect it over great distances. It is not necessary to postulate that a sexually mature female moth does any conscious thinking when she releases sex attractant into the air; this type of signaling can be explained as a direct result of physiological processes.

In many animals, however, reproductive communication is much more specialized. The female indicates by small changes in behavior that she is ready for mating, but her behavior is not merely a close correlation with the state of her reproductive system. Often there are complex sequences of communication between the female and one or more males as described by de Waal (1982) in chimpanzees. This means that the signal depends not only on the signaler's internal physiological state but also on the signals just received from the other animal. Through the exchange of signals, the communicating animals seem to learn something about the other's mood or likelihood to behave in certain ways.

These matters become clearer when one considers the competitive interactions that may occur among a number of males, each of which could mate with a given female. In many social animals the males attract the females rather than vice versa, and the ability to make conscious comparisons might facilitate the females' choices. In these species the males spend a great deal of time making displays of various sorts during the reproductive season. Females observe these displays before choosing one male, and mating occurs only after many further stages of courtship.

One of the best-known examples of male display is the territorial singing of male songbirds which produces such beautiful music to our ears (reviewed by Marler, 1977, 1978). Song has at least two biological functions. The male, having selected a territory, advertises to females his presence, his species, and his readiness to mate. His songs also broadcast to other males the message that he is ready to defend his territory. Males in adjoining territories sing loudly and often display to each other by assuming conspicuous and characteristic postures. Various kinds of aggression between males, much of it nonviolent, consisting of prolonged exchanges of displays or threatening gestures, lead eventually to the establishment of stable boundaries. Ritualized aggression does sometimes escalate to fighting and even to inflic-tion of serious injury. Ordinarily, however, after exchanging vocal or visual displays and threatening gestures, one male retreats, and his territory shrinks while that of his neighbor enlarges.

This outline, though sketchy, is sufficient to show the com-plexity of communication in the preliminaries to mating. More interesting still is the manner in which the female birds watch and listen to these displays and aggressive interactions between the males. It seems to be the rule rather than the exception for female birds approaching readiness for copulation and egg laying to watch and listen to numerous displaying males. This has led behavioral ecologists to the concept of female choice. One important aspect of evolution is sexual selection, whereby males whose displays lead females to select them as sexual partners leave more offspring than their less effective competitors. Sexual selection is believed to have resulted in some of the extreme anatomical structures and spectacular display behaviors found in animals, such as the peacock's tail and the elaborate songs of many birds. When such evolutionary developments go too far, however, they may inter-fere with efficient feeding or increase the risk of predation. Usu-ally a balance is reached in which male display behavior is effective in eliciting female choice but is not so great a handicap that the males fail to survive to reproduce.

These theories about sexual selection appeal to many behav-ioral ecologists and have been developed in elaborate detail, using refined mathematical formulations. But for our purposes, the im-portant point is that highly conspicuous structures for male ad-vertisement can have their evolutionary effects only if other members of the species react differentially to them. The females may or may not mate with a displaying male, and other males may retreat or attack and displace him from a preferred territory. Female choice was mentioned in Chapter 3 in connection with the idea that female blackbirds evaluate the territory held by a displaying male, but there are many indications that female choice is also influenced by the displays themselves. The two consider-ations must often be mutually reinforcing; vigorous males that win most of their competitions with other males are likely to have larger and superior territories, assuming that the males can also evaluate territories. Thus it is advantageous for females to select the males that have been most successful in their aggressive in-teractions with other males (reviewed by Green and Marler, 1979).

One aspect of these interactions leading to the choice of mates is important enough to mention here. In a few species of birds such as the sage grouse, as well as in a large African antelope, the Uganda kob, and a few other animals, the displaying males gather at particular locations which they use year after year, and display relatively close to one another. Such display areas are called leks; although the whole group may move together, as is the case with the Uganda kob, the lek always includes a central area where the older or more dominant males come day after day to spend most of their waking hours in vigorous display. They arrange themselves in smaller adjacent territories radiating outward from the central, dominant male. At the outer edges are the younger males, whose displays are clearly less vigorous. Females that are ready to mate come to the lek, move through the peripheral territories, and mate with the central male or one of his closest neighbors. This system leads to great disparity in reproductive success, with the central male fathering most of the progeny. Thus sexual selection operates in an unusually strenuous form in such birds as the sage grouse of the prairie states and its close relatives in other parts of the world (Wiley, 1974).

The pattern of ritualized aggression among competing males and simultaneous attraction of females is found in animals of many taxonomic groups. Among honeybees the males, called drones, congregate in the mating area and fly back and forth until an unmated queen flies to them. In the fantastically vigorous aerial interaction that ensues, one or more drones succeed in mating with the female. She then flies to her hive with her reproductive tract full of sperm, which last her for months or years.

Another dramatic example is provided by the fiddler crabs of the genus *Uca,* which are common inhabitants of mudflats and beaches in warm coastal areas. These crabs dig burrows in the sand or mud, to which they retreat during high tide. They feed at low tide on minute plants and animals in the mud, using the most anterior pair of legs to gather their food. In mature males, one anterior leg is greatly enlarged, and in many species rather conspicuously colored. This large claw is useless for gathering food, so the males suffer from a distinct disadvantage in feeding. The male crab uses this large claw almost exclusively for social display (Crane, 1975), waving it vigorously in patterns and rhythms characteristic of the species, as he stands beside his burrow. These waving displays are used both in aggressive interactions with other male fiddlers and in the attraction of females, though in somewhat different ways. When a female that is ready to mate comes within visual range, the male fiddler ordinarily stops displaying toward neighboring males and makes courtship displays toward the female.

Most of the male's time, however, is devoted to displaying toward other males. Often a wandering male approaches and challenges one that has been displaying beside a burrow. These challenges involve approach, pushing, and often interlocking the partly opened large claws, but they almost never lead to injury. This "hand shaking" seems to be important to male fiddler crabs, for their claws are so constructed that the tip of one male's claw rubs against a series of knobs and ridges on the other's claw. Since about half of the males have the right claw enlarged, and half the left, the interlocking bumps and ridges must be in different locations to accommodate like-handed and opposite-handed males. These mutual interlockings and rasping motions seem to determine the outcome of many of the ritualized fights, after which one crab retreats and leaves his rival at the burrow entrance. It would take us too far afield to describe the further complexities in the long-lasting and repeated aggressive displays.

To return to the general question, is there any reason to suppose that these communicative interactions are accompanied by conscious thoughts or feelings? It is now timely to recall the discussion in Chapter 2 of the idea advanced by Jolly (1966) and Humphrey (1978) that consciousness evolved in our ancestors because it was helpful, if not essential, for the successful cooperation of mutually interdependent creatures in social groups. We can ask whether in animals, also, the interactive cooperation required for female choice, on the one hand, and the effective synchronization of mating activities, on the other, might be facilitated by mutual awareness. In all the interactions discussed above, it seems important for a given animal to appreciate the mood and inclinations of other individuals. It may not be sufficient for one animal to treat another as belonging to one of a very few categories such as a female ready for mating or an aggressive male. The subtle complexities of these interactions suggest that finer-grain nuances may also be important. For example, when two males are engaging in elaborate and varied forms of ritualized aggression, each one must judge whether the other's threats are likely to escalate into injurious violence, and if they do, which animal is likely to prevail. So many elaborations on the general theme of threat, attack, and retreat would not seem necessary if animals were simply varying their aggressiveness along a linear scale of intensity. Extending Humphrey's suggestion might help us understand complexities in communicative behavior that otherwise seem to make no sense.

Can this rather vague speculation be rendered more explicit? Is it possible that in the many subtle variations on the basic themes of aggression and displays to females, animals are communicating and judging each other's moods, behavioral dispositions, even thoughts and feelings, better than they could otherwise? The relatively simple exchanges discussed so far do not answer this question, but in certain cases we know enough about the behavior and its effects on social companions to venture a little further.

One such case involves the complex territorial songs of certain bird species. While many species sing simple, repetitive patterns, others—for example song sparrows, certain wrens, and domesticated canaries—produce elaborate songs. If the function of a male's song is simply to announce his presence and advertise his vigor and success, and to strengthen his threat against other males, why should such elaborate songs have arisen? An indication of the possible function of song variety is shown in the results of experiments by Kroodsma (1976). When he played tape recordings of various canary songs to female canaries that were coming into breeding condition, he found that the more varied songs stimulated the females more strongly, as reflected in the amount of nest building and other behavior normally associated with preparation for mating and raising young. Thus song variety may be one more instance where sexual selection through female choice has enhanced a behavioral attribute that would otherwise have no particular function or survival value. But why should more varied songs make a male more attractive? Perhaps the elaborateness of a song conveys subtle and varied messages either of threats against neighboring males or of sexual ardor for the female.

Communication as a Window

These considerations lead us back to some basic points about human communication and language. All of the known human languages are much more versatile and complex than any known animal communication system, but language is not our only means of communicating. Scientists have come to recognize the large role played by nonverbal communication, which includes gestures, facial expressions, bodily postures, gait in walking, nonverbal sounds, eye contact, and the like. The messages tend to be rather general and often convey emotional states rather than specific information about discrete objects, but a pointing gesture, for example, can be very specific. One attractive line of thought is to liken animal communication to the nonverbal communication of people, while holding to the conviction that language is a unique human accomplishment. But as I shall argue later, this distinction may not be absolute.

This leads us back to a basic philosophical idea about the nature of the human mind. When confronted by the solipsist's logically consistent but highly implausible argument that he is the only conscious thinking creature in the universe, we are obliged to ask how any of us really knows that other people are also conscious and experience subjective feelings. But although rigorous proof is not logically possible, we do have abundant reason to believe that our fellow human beings are also conscious creatures, who experience a variety of subjective feelings and whose mental experience is not greatly different from our own.

We are convinced that our companions experience feelings and thoughts primarily because they tell us about them. As civilized adult thinkers we convey specific information about our thoughts, mainly through language, although nonverbal signals also play a large role. Our best—perhaps virtually our only—way of learning about other people's thoughts and feelings is through verbal and nonverbal communication.

From these elementary considerations there emerges a very simple but potentially significant point, which I have discussed elsewhere in some detail (Griffin, 1978, 1981). If nonhuman animals experience conscious thoughts or subjective feelings, we might be able to learn about them by intercepting the signals by which they communicate these thoughts and feelings to other animals. This idea is so basic that scientists, accustomed to dealing with complex issues, find it difficult to appreciate its potential significance. Nevertheless, the analogy to how we learn about other people's thoughts and feelings is so directly appropriate that we should make an effort to see where it might lead.

Two objections have been advanced against accepting communication as a useful window on animal thinking. The first stems from the belief that animals communicate only their own emotional states, such as fear, aggressiveness, sexual motivation, hunger, generalized distress, or other undifferentiated internal states. We are already used to interpreting yelps of pain, threatening growls, or distress calls; indeed, all who deal with animals are used to intercepting these emotional signals. Only by denying that animals have subjective experiences such as pain, hunger, distress, or sexual urges is it possible to deny that a person can learn about an animal's emotional state from observing its signals to another animal.

Another version of this negativity is the assertion that the sounds, expressions, postures, or other signals an animal displays when it is hungry, threatened, or fearful are merely incidental byproducts of its physiological functions during these emotional states. But in many cases it is clear that other animals obtain information from emotional signals directed specifically at them; rivals retreat before threats, parents bring food to their young upon hearing their hunger calls, and distress calls very commonly bring a parent rushing to its young. Communicative signals are seldom produced at all unless an appropriate receiver is present, usually a companion that has a specific social relationship to the communicator. If the signals were only automatic byproducts of internal states, such as groans of pain, one would expect their occurrence to depend only on the animal's internal state, not on the presence of a responsive receiver.

The major area of disagreement concerns whether we can learn something more specific about the content of animal thoughts by studying their communication. An emotional signal may tell us that the animal is afraid, but can it tell us what it is afraid of? Many ethologists agree with Smith (1977) in denying that animals, other than perhaps the great apes, dogs, and porpoises, are capable of thinking about objects or events. They generally assume that animals react unconsciously and perhaps feel basic emotions, but are unable to picture anything other than immediate sensations. They are held to be incapable of considering alternatives in deciding what to do. Can animal communication provide any information bearing on these beliefs? Do animals ever communicate about specific things? When young animals beg their parents for food, they are certainly conveying that they are hungry, but do they ever ask for a particular kind of food? Ethologists seldom ask such questions, but young birds and mammals often do seem to prefer certain types of food, at least in the late stages of being fed by their parents. Unfortunately, we have very little data from which we can judge whether food begging calls convey information about specific kinds of foods.

Parrots are generally believed to mimic human speech without the slightest understanding of the words they imitate. But Pepperberg (1981, 1983) has succeeded in training an African grey parrot named Alex to name and request up to about forty different objects by means of clearly understandable English words. His vocabulary includes names of familiar objects and actions, as well as a few adjectives describing color or shape. Alex can correctly name objects shown to him; he also requests specific objects and plays with them when provided by the experimenter. When given a different object from what he asked for he protests with a loud "no." These discoveries are so surprising to behavioral scientists that they have not yet been replicated and taken seriously. But they certainly call into question the limitations we have customarily placed on the communicative and cognitive capabilities of birds.

Some scientists have claimed that animal communication is no more promising than many other categories of behavior as a source of evidence about their thinking. The many kinds of laboratory experiments on learning and problem solving discussed in Chapter 7 are held by these scientists to be a form of communication with the animal superior to the signals it naturally exchanges with its companions. In this view the experimenter, by placing the animal in the experimental situation, is asking a question, and the animal, through its success or failure in solving the problem, is giving an answer. But this very limited form of communication tells us primarily what the animal can or cannot learn to do and does not reveal much about what it may think or feel. The study of specialized communicative behavior between animals is more promising than learning experiments because it seems to have evolved to serve a purpose quite similar to what we want to use it for—namely, conveying to another animal the content of the communicator's thoughts and feelings. If its thoughts do have significant content, eavesdropping on its communicative behavior under favorable conditions might tell us something about that content.

According to many ethologists, animal communication signals seem to lack information about objects or events other than those occurring within the communicator. Human language, on the other hand, is extremely versatile in this respect, and we can describe in considerable detail real or imagined objects or events that are remote in time or space from the immediate situation. To the degree that socially interacting animals need be concerned only with the moods or dispositions of their companions, communication about oneself or one's immediate situation may suffice.

For example, the specialized distress calls of baby mice convey to the mother not only that her infants are distressed, but also information about the temperature in their nest. However, the message concerns their own bodily state, not a general description of air temperature (Sales and Pye, 1974). Most instances of animal communication seem to relate only to the communicator's immediate situation here and now.

But some social animals need information about more than the immediate situation or the communicator's internal state. It may be important for cooperative animals to be informed about dangers, food supplies, or other important matters, even when these are not part of any one individual's here-and-now. Such situations call for more versatile communication, and possibly for more versatile thinking. Students of human language and of communication theory will realize that I am talking about the property called "displacement," meaning that the information refers to something displaced from the immediate situation. Many have argued that displacement is never, or only very rarely, an attribute of animal communication, but in the next chapter I will describe two cases where animal communication about matters of interest to the whole group does have the property of displacement.

$\mathcal{9}$ Symbolic Communication

One recently studied example of communication conveying more than the internal state of the communicators involves African vervet monkeys. These monkeys, about the size of a small dog, have been studied extensively by ethologists because they are abundant in open areas where much of their activity can be observed, and because they live in quite stable social groups. Each group is closely interrelated, consisting of a few generations of parents, offspring, cousins, and other close relatives. Inbreeding is reduced by migration of young males to neighboring groups from time to time.

By close and careful observation, ethologists have learned to recognize the individuals in these groups and have been able to study the behavior of each one relative to his companions over extended periods. Learning to recognize individual animals under natural conditions, a very important advance in ethology, has led to the discovery of previously unsuspected patterns of behavior that could not be appreciated when the animals were treated as interchangeable units. One discovery has been that not only the ethologist but the animals themselves often recognize their companions as individuals and treat them accordingly. In many primate societies, dominance depends to a considerable degree on who your parents are. A young monkey that could not successfully threaten another member of the group may be given access to food because it is the offspring of a dominant mother who can be expected to intervene in any aggressive encounters that may develop.

Semantic Alarm Calls

Vervet monkeys have at least three different categories of alarm calls, which were described by Struhsaker (1967) after extensive periods of observation. He found that when a leopard or other large carnivorous mammal approached, the monkeys gave one type of alarm call; quite a different call was used at the sight of a martial eagle, one of the few flying predators that does capture vervet monkeys. A third type of alarm call was given when a large snake approached the group. This degree of differentiation of alarm calls is not unique, although it has been described in only a few kinds of animals. For example, ground squirrels of western North America use different types of calls when frightened by a ground predator or by a predatory bird such as a hawk (Owings and Leger, 1980), but some ethologists feel that the difference relates to how frightened the squirrel is or how far it is from its burrow. When ethologists interpret data of this kind, they tend to search hard for interpretations that avoid any hint that animal communication signals convey anything more than information about the communicator's internal state.

The first and relatively simple question is whether the vervet monkey's three types of alarm calls convey to other monkeys information about the type of predator. Such information is important because the animal's defensive tactics are different in the three cases. When a leopard or other large carnivore approaches, the monkeys climb into trees. But leopards are good climbers, so the monkeys can escape them only by climbing out onto the smallest branches, which are too weak to support a leopard. When the monkeys see a martial eagle, they move into thick vegetation close to a tree trunk or at ground level. Thus the tactics that help them escape from a leopard make them highly vulnerable to a martial eagle, and vice versa. In response to the threat of a large snake they stand on their hind legs and look around to locate the snake, then simply move away from it, either along the ground or by climbing into a tree.

Knowing that the monkeys give different alarm calls when they see different predators does not establish beyond doubt that the calls actually describe the type of predator. When the monkeys, which are usually close to each other, hear an alarm call, each one quickly looks around at the caller. Like many other animals, they are adept at judging the direction in which another animal is looking, so they can easily see what the caller is looking at. This serves much the same function as pointing. When other monkeys than the caller take the appropriate action to avoid the danger, it is difficult to be sure whether they are acting solely on the basis of the call or whether the call simply led them to look at the source of danger.

In addition, many ethologists have argued that the different alarm calls convey only degrees of fear, that they reflect a scale of intensity rather than information about the predator. Although the three alarm calls sound quite different, and although all three can vary in loudness, many ethologists have convinced themselves that they constitute nothing more than steps along a scale of intensity of fear.

To clarify this situation, Robert Seyfarth, Dorothy Cheney Seyfarth, and Peter Marler (1980a, 1980b) conducted some carefully controlled playback experiments under natural conditions in East Africa. The basic idea was to play from a concealed loudspeaker tape recordings of vervet alarm calls when they had just seen a leopard, a martial eagle, or a large python, and to inquire whether these playbacks, in the absence of a predator, would elicit the normal response. The experiment required many precautions and refinements. For instance, vervet monkeys come to know each other as individuals, not only by visual appearance but by minor differences in their vocalizations. They might not respond to recordings of alarm calls from another group, and they might not respond even to an alarm call recorded from one of their own companions if that individual was in plain sight some distance from the vegetation concealing the speaker. In all experiments the loudspeaker reproduced calls of a member of the group, and the speaker was hidden in a place where the monkeys would expect that individual to be. The experimenters had to be prepared with tape recordings of a known member of a well-studied group of vervet monkeys, and to locate the hidden speaker where this individual frequently spent time. The other monkeys had to be observed when they were not actively engaged in some other behavior and were not reacting to another danger.

When all of these conditions were satisfied, the playbacks of alarm calls did indeed elicit the appropriate responses. The monkeys responded to the leopard alarm call by climbing into the nearest tree; the martial eagle alarm caused them to dive into thick vegetation; and the python alarm produced the typical behavior of standing on the hind legs and looking all around for the nonexistent snake. In this brief description I have omitted a number of technical details and controls that were needed to assure that the data were correctly interpreted. For example, along with evaluations by the ethologists on the scene, sound movies of the behavior were interpreted by scientists who did not know what type of call had been played back.

Not all ethologists have accepted the straightforward interpretation that the alarm calls convey information about the type of predator. One alternative interpretation is that the alarm calls are injunctions to behave in certain ways. Thus the leopard alarm might mean "Go climb into a tree." But even this interpretation, which seems somewhat strained, necessarily ascribes three specific types of injunction to the vocabulary of vervet monkeys. Even such postulated injunctions would be more than a simple reflection of the internal state of the communicator—unless one begs the question by so extending the definition of internal states as to include the impulse to give out differentiated messages.

The Seyfarths' further studies of vervet monkeys have disclosed numerous other subtleties of social communication, of which I will briefly mention two. In some experiments the screams of a particular juvenile monkey were played back from a hidden loudspeaker at a time when he had moved out of sight in that general direction. Not only did his mother show strong responses, but other adult females looked not at the loudspeaker, but at the mother of the youngster whose screams were being reproduced. From the sound, they knew the identity of both the young monkey and its mother.

More recent studies have analyzed the relatively low intensity and highly varied grunting sounds of vervet monkeys engaged in amicable interactions such as mutual grooming. Cheney and Seyfarth (1982a) have found evidence that vervet monkeys distinguish among four types of grunts that are very difficult for human listeners to tell apart. They use the four varieties in different social contexts, such as when approaching a dominant or a subordinate companion, when moving into a new area within the group's familiar range, or when first noticing another group of vervet monkeys. Like the alarm calls, the grunts function as though each conveys a specific meaning.

Although a great deal remains to be learned about the subtleties of vocal communication by vervet monkeys and other primates, these revolutionary discoveries should warn us that it has been a serious error to assume that monkeys exchange only crude messages. The versatility of animal communication may have been overlooked because students of animal behavior have been so thoroughly convinced that it was unthinkable. Ethologists should now be alert to the possibility that semantic communication occurs in a wide variety of social animals.

Communication about Distant Food or Battle

The division of labor in large social groups, of animals or people, usually increases the efficiency of the group. By having some group members concentrate on one essential activity and some on another, the effectiveness of the whole group is often enhanced. In many species food is gathered through some division of labor, with certain members searching for food at a considerable distance, gathering more than they need themselves, and bringing back the rest for their companions.

On a very small scale, nesting birds do this when they gather food at some distance to feed their young. In a few bird species one parent may remain with the young while the other gathers food for both mate and young. An extreme case of division of labor is that of the hornbills of tropical Africa and Asia. These large birds build their nests in tree cavities. After the female has completed a nest, the male closes most of the entrance to the cavity, locking the female inside. He then passes food to her through the small hole that has been left open. This confinement, which probably protects mother and young from predators, continues for several weeks while the young develop; finally the mother breaks down the wall and begins to help collect food for the young. The young themselves close up the cavity entrance after the mother has departed, and for an additional period they are fed by both parents (Frisch, 1974).

In practically all birds, feeding the dependent young places great demands on the parents' energies. They must gather enormous amounts of food, and this may require prolonged searching over a wide area. When both parents are strenuously engaged in this task, they would benefit from being able to pool information about food sources. But ethologists have not observed communication between parent birds about the location of food, even in species whose efficiency would clearly be enhanced. Many species of insect-eating birds must hunt over wide areas to find enough insects. When they find a good supply, repeated trips to the same area are likely to be more rewarding than searching elsewhere, but to the best of our knowledge birds do not communicate such information to their mates.

But in other, distantly related animals, communication about the location of food sources has been described in great detail. We usually feel that the most complex and versatile sorts of animal behavior will be found in our closest animal relatives, or at least in animals that are similar to ourselves in brain size and shape, a viewpoint emphasized by Walker (1983). Thus we are not unduly surprised to find chimpanzees and also dolphins which have brains as large and as complex as ours, exhibiting versatile, even insightful, behavior. But when it comes to social behavior, division of labor, and mutual interdependence, the most striking examples are found not among mammals, or even vertebrates, but among the social insects.

If we can restrain our initial gut reactions that no mere insects could be capable of thoughtful behavior, we will find several surprisingly intricate and suggestive examples of social communication among ants and bees. We must face at the outset the strong and very widespread belief that no matter how complex and effective an insect's behavior is, it is too small and its central nervous system too differently organized from ours to be capable of conscious thinking and planning or subjective feelings. Recognizing this deep-seated belief, let us consider what certain social insects actually do.

The first example is the recruitment behavior of weaver ants, remarkably efficient builders of leaf nests described in Chapter 5. As in other large colonies of social insects there is considerable division of labor. Some of the nonreproductive workers gather food and bring it back to the several nests, while others feed the queen and the larvae, clean the nest, repair damage to it, and fight off other insects that threaten to intrude. The most interesting communicative behavior is exhibited by the workers that forage at some distance from the nests (Hölldobler and Wilson, 1978). When they have found a good source of food, they return to the nest and engage in behavior that serves to recruit other workers to join them in gathering the food. As in many other species of ants, they do this in part by laying odor trails along the route from the nest to the food source (Wilson, 1971). But they also must induce additional workers to help them carry away the food before other animals get to it.

Hölldobler and Wilson have studied what the weaver ant workers actually do when they recruit their sisters to gather food or to participate in some other activity at a distance from the nest. The recruiter engages in a series of face-to-face encounters with other workers in which they feel each other with their antennae, which probably transmits the odor of the food. The recruiter also makes lateral head wagging movements. This often, though not invariably, induces the other ant to follow the odor trail to the food source. This type of recruitment of workers is not radically different from the comparable behavior of many other ants, but the lateral head movements seem to be particularly intense and consistent.

These ants use a different set of recruiting gestures when they are attacked by other insects. Such intruders may overrun the colony, killing and eating adults and young, so their approach in numbers is a threat of the first magnitude. When some of the foraging workers from a colony of weaver ants encounter intruders, even at a considerable distance from the leaf nest, they stop searching for food and begin to fight and also release communicative alarm pheromones. If the intruders are numerous, some workers break off fighting and return to the nest, laying an odor trail similar to the trails used to mark a route to a food source, although we do not know whether the odorous substances are the same. Back in the nest they recruit other workers by jerking their bodies back and forth toward and away from the nestmate rather than side to side as in recruitment for food gathering. According to Hölldobler and Wilson, weaver ants communicate in other ways also, such as exhorting their nestmates to move to another location, but the distinction between gestures is especially clear in the case of food sources and intruders.

Hölldobler and Wilson limit themselves to commenting that the recruiting gestures used in response to dangerous intruders resemble in some ways the movements employed in fighting. They interpret this as a sort of ritualized version of actual fighting, but it can also be interpreted as intentional pantomine. In any event, the recruitment is effective, and many workers rapidly move out to combat the intruders. Why should recruitment for these two needs involve different gestures? It seems possible that each type of communication conveys to the recruited ants a message about what is to be done at the end of the odor trail. Perhaps they are more effective fighters if they are prepared for fighting rather than for food gathering. Might such preparation include a simple thought such as "Now we fight"?

The recruitment of weaver ant workers to join in fighting includes another feature that is very rare among insects, but highly suggestive as evidence that the communication reflects thinking. Some of the ants on the receiving end of the recruitment gestures, instead of immediately following the odor trail, turn to other workers and repeat the same recruiting gestures, even though they have not been directly stimulated by intruders. This sort of chain communication is very effective in rapidly enlisting a large number of fellow fighters. But it also exhibits a property not ordinarily found in animal communication, namely, the conveying of specific information about something the communicator has not been exposed to directly, but has learned about only by receiving communicative signals. To be sure, an animal that has been aroused by an alarm call will often repeat it, but these repeated calls are believed to be nonspecific and to convey only the communicator's state of alarm. Here there is at least some specificity in the distinction between recruitment gestures meaning "Go get food" and those meaning "Go fight."

Could ants communicating with their sister nestmates be consciously aware of the simple messages Hölldobler and Wilson have decoded? Recognizing that most communication between ants involves chemical signals, which would be almost impossible to decode if they are at all differentiated, we should maintain open minds about the possibility that these simple distinctions among recruitment messages might be only the more easily detected components of a complex and versatile system of chemical communication. I am extrapolating here far beyond anything that these scientists or their colleagues have suggested. When asked how they know that these ants cannot entertain any conscious thoughts, the scientists usually fall back on the prior conviction that insects are genetically programmed automata and that their brains are too small to permit conscious thinking. Evidence of differentiated communication has not yet altered this deep-seated conviction.

The central nervous system of ants are minute compared to those of even the smallest birds and mammals. But how can we be certain about the critical size necessary for conscious thinking? Even the smallest insect brain contains thousands of neurons, each capable of exchanging nerve impulses with dozens of others.

The content and complexity of conscious thoughts, and their range of versatility, might be roughly proportional to the volume of the central nervous system, but an absolute criterion of mass necessary for conscious thinking is not supported by anything we know about the nature and functioning of central nervous systems.

A general belief among biologists that has been taken over without question by psychologists and philosophers is that only the brains of vertebrate animals are large and complex enough to permit versatile behavior or conscious thinking. Some grudgingly accept that the obviously complex brains of octopus and other large cephalopods are also capable of a certain degree of flexibility. A corollary of this dogma is that only a central nervous system that is concentrated into a dorsal nerve cord, preferably enlarged at the anterior end, can support anything other than the most stereotyped reflexes. For this reason alone, arthropods, with much but not all of their central nervous tissue located in paired ganglia ventral to the digestive tract, are relegated to the status of genetically programmed automata. Explicit statements of this belief have recently been advanced by Grene (1978) and, in a more thoughtfully tempered fashion, by Walker (1983). For the most part, the lower vertebrates and invertebrates are simply ignored in Walker's review of behavior suggesting animal thought. He repeatedly expresses surprise, bordering sometimes on shock, as he reviews evidence that a fish or—horrors—an insect can learn and adapt its behavior to changing circumstances.

What underlies this dogma that only a vertebrate central nervous system is capable of organizing thoughts? Historically this belief was based on finding that animals whose behavior seemed flexible and adaptable had the vertebrate pattern of gross neuroanatomy. But how did we come to be so certain that only vertebrate central nervous systems are capable of learning or versatile thinking? The chief basis for inferring that insect thinking is very limited has been evidence of behavioral fixity and limited versatility. But new discoveries, such as those of Hölldobler and Wilson and many others reviewed by Lindauer (1974), Markl (1974), and Lloyd (1981a, 1981b, 1983), have shown that the behavior of some insects is far more flexible and versatile than previously recognized. Furthermore, learning is now known to be well within the capabilities of many annelids, mollusks and arthropods. Perhaps this new behavioral evidence will modify our long-standing conviction that all invertebrates are thoughtless automata.

At least a century has elapsed since this generalization based on the gross anatomy of central nervous systems was accepted. In the meantime much has been learned about how brains function, and nothing that has been discovered suggests significant phylogenetic differences in the basic mechanisms. Everyone agrees that it is the patterns of organization of neurons and synapses that allow brains to function as they do. But the neurons need not be clumped in any particular shape or other grossly visible structure. To be sure, it is important for clinicians and investigators to know just where in the vertebrate brain certain functions are concentrated, but there is no basic principle that requires, for example, chemical sensitivity to be handled at the front of a brain, or posture and balance toward the rear.

It is thus wholly unconvincing to argue, as we have all been trained to do, that the ladderlike arrangement of paired ganglia in an arthropod's central nervous system precludes flexible behavior or conscious thinking. Arthropods also have brains where sensory information, especially from the eyes and the tactile and chemical receptors on the antennae, is received and integrated. Neurobiologists have analyzed many mechanisms of information processing in the central nervous systems of insects and other relatively complex invertebrates. Cells in one area inhibit activity in other regions, and both sensory input and motor output are regulated to produce versatile behavior (Bullock and Horridge, 1965; Treherne, 1974). So far, only relatively simple central integration processes have been analyzed in detail (Huber, 1974; Hoyle, 1977; Pearson, 1977; O'Shea and Rowell, 1977; Burrows, 1977, 1982; Elsner and Popov, 1978). But experimenters have demonstrated that the central nervous systems of crustaceans, insects, and cephalopods organize and modulate information in ways that are quite comparable in complexity and precision to those of vertebrate brains. Hence we are forced to fall back on the small size of invertebrates' central nervous systems as a reason for doubting their versatility. The complexity of an animal's behavior and thinking may be proportional to brain volume, but that does not mean that there necessarily is a qualitative difference in kind between vertebrates and invertebrates.

The considerations developed in the final section of Chapter 2 are pertinent here. If we examine the assumption that genetically programmed behavior cannot be accompanied by conscious thinking, we find that it is based on a tenuous analogy to those few patterns of human behavior that seem likely to be based on genetic information rather than individual experience. Because of the supposedly great differences between our own and other species, scientists frown upon reasoning by analogy from human conscious thoughts and subjective feelings to those of other animals under similar conditions, but no direct evidence supports the restriction of conscious thinking to a particular form of gross neuroanatomy.

The complex societies of ants are enormously impressive in their specialized division of labor. In addition to digging burrows, laying eggs, tending developing larvae, and mating, some ant species literally domesticate insects from entirely different taxonomic groups, such as aphids. Many of these relationships are symbiotic in the sense that both species clearly benefit. The ants protect the aphids, even build shelters for them, and in return obtain food that would be difficult or impossible to obtain otherwise. The variety of these arrangements and of the structures in which the aphids are housed is extremely impressive, as documented in *The Insect Societies* by E. O. Wilson (1971) and the four-volume *Social Insects,* edited by Hermann (1979–1982).

A frustrating technical problem in learning about communication in ants and most other social insects is that almost all of it is done through exchange of chemical signals. Specialized molecules produced in a variety of glands are emitted at particular times to stimulate other individuals, either in tactile contact at close range or at a distance, as in the case of sex attractants. Substances that convey information over long distances are difficult to study because they are effective at very low concentrations, but several have been chemically identified and it has been confirmed that they attract males even in the absence of a female. It is much more difficult to study the chemical substances used for close-range communication through contact or the exchange of material from stomachs. The effective quantities are much smaller, because there is no loss in transmission from one insect to another. Furthermore, there are many indications that unlike the sex attractants, these chemical signals consist, not of a single active substance but of several (Bradshaw, 1981; Bradshaw, Baker, and Howse, 1975, 1979; Bradshaw et al., 1979).

The only methods currently available for studying these chemical contact signals are rather crude, such as crushing entire glands and presenting their juices to other insects, yet even these have been revealing. For example, some species of ants recruit nestmates to food sources by what is called tandem running. After antennal contact the recruited ant grasps the abdomen of the recruiter and the two run in tandem from the nest to the source of food. If a crude extract from the glands of the recruiter is smeared on the end of a tiny stick, a recruited ant will often hold on to the scented stick and run in much the same way as with a recruiter ant. Such experiments help build up the picture of insects as genetically programmed robots, for one would suppose that a more thoughtful creature would realize that the stick was not really a sister worker.

Social insects do not always respond in the same way after exchanging contact chemical signals with nestmates, and it is therefore possible that such chemical signals may convey more complex and versatile messages than anything yet decoded by ethologists. But more refined experiments are very difficult to design. Ideally we should determine what the active substances are, the concentrations in which they occur, and the different signals conveyed by different amounts or combinations. This would allow experimenters to try reproducing definite behavior patterns by delivering particular chemical signals. Because such investigations have not been practicable, we do not know whether chemical signals convey more complex messages than generalized arousal or the injunction to follow an odor trail. The tactile signals used in recruitment by weaver ants can be observed and their nature correlated with the resulting behavior, making the study of communication by gestures far more feasible than the study of chemical communication.

Symbolic Gestures

Another example of versatile communication among social insects, involving both chemical and tactile signals, has yielded far more extensive evidence with even more significant implications. This is the so-called "dance language" of honeybees, which was discovered by the remarkably brilliant and original experiments of Karl von Frisch (reviewed by Frisch, 1967, 1972). These communicative dances are so astonishingly different from other animal communication that they seem discordant with everything else we know about animal behavior, to the point that they have not been adequately integrated into scientists' understanding of ethology. One behavioral ecologist has called them an "evolutionary freak" (Krebs, 1977). However, comparable, but simpler, communicative gestures are known in other species of bees, and Martin Lindauer (1971) has traced a probable sequence of evolutionary stages from patterns of behavior found in many kinds of insects to the remarkably versatile communcation in honeybees.

The most elaborate form of this versatile type of communication is the so-called waggle dance or *Schwänzeltanz*. A worker honeybee crawls rapidly over the surface of the honeycomb in a flattened figure-eight pattern, ordinarily followed by several other worker bees that press their heads and antennae against her body. The central portion of the figure eight is nearly straight, followed by circular return movements alternately to the right and the left, each of which brings the bee back to the starting point. The straight run is the important part of the dance; the circular return movements seem just to get the bee in position to repeat it. During the straight run the dancing bee lashes her abdomen from side to side approximately thirteen times per second. During this vigorous motion the bees pressing close against her body are strongly stimulated, both mechanically and chemically by numerous scents transmitted from the dancer to the sensitive chemical sense organs on the antennae of the other bees.

Waggle dances are most often employed when the older worker bees that do the food gathering have located a rich source of food and brought back some to the colony. But to elicit a waggle dance, that type of food must be in short supply within the colony. American beekeepers have remarked that they seldom see waggle dances, but this is because their bees usually have access to abundant food. Only when food or some other important commodity is badly needed by the colony are these communicative dances performed. How can a worker honeybee find out that something is needed by the colony? This important point is often overlooked in elementary descriptions of the bee dances. They do not occur in isolation, but are part of a complex system of social communication, most of which involves both chemical and tactile signals.

Worker honeybees spend a great deal of time moving about in the hive and interacting with each other by antennal feeling and, in many cases, also by regurgitating stomach contents, which are then taken into the stomach of another bee. This sort of exchange, widespread among social insects, is called trophallaxis. It may offend our aesthetic sensitivities, but it is an important means of social communication that enables insects to coordinate their activities. Many members of a social insect colony are fed in this way, including the queen and the numerous larvae that could not possibly leave the colony and gather their own food. But similar food exchange occurs among workers that are quite capable of flying outside the hive but do not do so; instead they clean the hive and pass on to larvae food they have received through trophallactic exchange with older workers returning with nectar or pollen from distant flowers. Trophallaxis causes food and odors associated with it to be widely distributed throughout a colony; a given sugar molecule may pass through several stomachs before becoming part of the honey stored in cells of the honeycomb. Pollen, brought back in specialized pollen baskets formed from stiff hairs on the legs, is also transferred by other bees before storage. A surprising proportion of the time the workers seem to be doing nothing at all (Lindauer, 1971), but perhaps even when "loafing" they are receiving information about the conditions inside the hive.

Honeybees also gather waxy and resinous materials when needed for construction and repair of the hive, and in hot weather they bring in water to cool the hive by evaporation. A forager that has taken up water in her stomach regurgitates it, often on the surface of cells containing growing larvae; then other workers spread it into a thin film over the larval cell, while still others fan vigorously with their wings, creating air currents that speed the evaporation of the water and cool the immediate vicinity. Worker bees that have been gathering food or other material outside of the hive are informed about what materials are needed when they transfer their load to other workers. If the hive has suddenly become overheated, workers returning with pollen or concentrated sugar solution find it much more difficult to unload their stomach contents or pollen baskets. As a result of this communication network, which I have described only in barest outline, those worker bees that are foraging outside the hive adjust what they are searching for—that is, their searching images—and hunt for what they have learned is needed (Lindauer, 1971).

This sounds like rather complex behavior for genetically programmed robots. But we are asked to believe that part of the instruction from the DNA of honeybees leads, through some unknown series of developmental processes, to behavior organized in such a way that when they have difficulty unloading one type of material they automatically search for something else.

The waggle dances are carried out only under certain social conditions. A particular worker has received the information through contacts with several of her sisters that sugar (or something else) is in very short supply. She must then have flown out, searched for flowers, and located a good source of flowers containing nectar with a high sugar concentration. There is a further differentation in her behavior depending upon the approximate distance of the food source. If it is within 30–50 meters of the hive, her dance consists of circles, alternating first in one direction and then the other without the straight waggling run described earlier. This is called a round dance. As the distance increases a brief waggle is inserted between the clockwise and counterclockwise circles, and the duration of this phase gradually increases with the distance to the food source. The distance or the effort required to fly to the food source is most clearly represented by the duration of the waggling run. But since the waggling motion is relatively constant in its details, a longer duration is almost always accompanied by a longer distance over the surface of the honeycomb.

The waggle dance conveys two other important kinds of information. The most astonishing of von Frisch's discoveries is that the dance tells the direction to the food source in relation to the position of the sun. A waggle run pointed straight up on the vertical surface of the honeycomb means food or other desirable material in the direction of the sun, straight down means away from the sun, and so on at intermediate angles. Waggle dances sometimes take place on a horizontal surface near the entrance to the hive, where the sun is visible. On a horizontal surface the direction toward food is indicated directly rather than being transferred to a vertical angle relative to straight up. In this condensed and oversimplified description I have omitted several important details, but let us think about what even this much implies. A worker bee executes inside the pitch-dark hive a geometrical pattern of motion which encodes both the distance and direction to a source of food or, as explained below, of something else. Assuming that when a few of her sisters follow very actively in close proximity to the dancer they correctly interpret this coded information, we have a communication system that is truly symbolic and that also displays in explicit form the property of displacement. Gestural motions inside a dark hive, which one bee detects by feeling the motions of another, stand for direction and distance to be flown outside the hive. Furthermore, the critical reference point for direction is the position of the sun in the sky, which is completely invisible within the dark hive. The upward direction has come to stand for direction relative to the sun.

The waggle dance conveys a second important message by its

vigor or intensity, primarily through the amplitude of the lateral motions of the abdomen (the frequency is approximately constant). Experienced human observers watching through the glass window of an observation hive can readily judge whether the dances are vigorous and emphatic or feeble and unimpressive. The vigor of the waggle dance varies with the desirability of the food, as measured by the concentration of sugar in the nectar on which the bees are feeding or the colony's need for something other than sugar. A dance about water can be very vigorous if the hive is overheated. The workers that cluster about a dancer respond far more actively to vigorous waggle dances. When the dance is not vigorous, very few bees go to the food source indicated. The waggle dance communicates three important attributes of a distant object: direction, distance, and desirability. Is this too much to attribute to a genetically programmed robot?

In a long series of experiments von Frisch inquired how much information about distance and direction is actually conveyed. When many bees were visiting one feeder it was removed, and several identical test feeders were set out at various distances and directions from the hive. The number of bees arriving at these feeders supported the conclusion that information about distance and directions is indeed transmitted, because more newly recruited bees arrived at test feeders situated at approximately the same distance and direction as the original feeder from which the dancers had returned a short time previously. But it was later pointed out that odors rather than the dance might have guided the bees. Several years earlier, von Frisch had conducted intensive studies of bees' chemical senses and their ability to distinguish among the odors of different plants. His experiments had shown that information about rich food sources is transmitted by plant odors either in pollen grains adhering to the dancer's body or in the nectar which has been regurgitated and taken up by other bees. While the dances convey the approximate distance and direction, they are only accurate enough to bring recruits to the general vicinity of the food source. The flowers must then be located by scent.

James L. Gould has since conducted more elaborate, carefully controlled experiments in which he caused a dancer to point her waggle run in a direction different from that of the food source from which she had returned. His test feeders were constructed so as to preclude a variety of complicating effects not adequately eliminated in von Frisch's experiments. The results showed that most of the recruited workers did indeed search for food in the direction indicated by the dance rather than the direction from which the dancer had returned (Gould, 1975, 1976; reviewed by Griffin, 1981). There now remains no serious doubt that information about distance, direction, and desirability of food sources really is conveyed from one worker bee to her sisters through these symbolic dances. This does not mean that the communication system is flawless and that every bee obediently flies precisely where directed. There is much variability not only in the dances but in the bees' responses. Odors also form an important part of the communication system, but the symbolism used in transmitting distance and direction effectively enables bees to exploit new food sources many hundreds of meters from the hive.

Several interpretations of these symbolic dances have been advanced in an effort to avoid the sort of cognitive interpretation I have expressed (Gould, 1979). One of the first "nothing but" interpretations claimed that dancing was rigidly linked to food and that the wagging of the abdomen was related to the sugar concentration in the dancer's stomach or blood. But almost from the start, von Frisch emphasized that the same type of dance communication is also used when the colony is in serious need of quite different materials, such as pollen as a source of protein, water to cool an overheated hive, or materials for repairing parts of the hive.

The most important extension of von Frisch's discoveries was made by Martin Lindauer (1955), whose truly classic experiments have never been adequately appreciated. When a colony of honeybees increases in number to the point that the hive is not big enough, new queens are produced; workers cause this to happen by feeding different food to a few of the larvae. Under ordinary conditions the bees also prepare to swarm, indicated by a change in the behavior of the older worker bees that have been gathering food. As new queens develop, the original queen stops laying eggs and moves out of the hive along with a large fraction, commonly half or more, of the workers. These form an impressive swarm, a ball of bees clinging to the surface of the hive or to some vegetation nearby. In agricultural practice, the beekeeper either enlarges the hive at the first signs of swarming so that the colony can grow further, or he provides a new empty hive immediately below the swarm, which the bees usually move into and found a new colony.

But when no beekeeper intervenes, this naked ball of bees

urgently requires a new cavity where the colony can be reestablished and continue its existence. The older workers that have been searching for food begin to search for another cavity with the right properties to serve as a new home. It is important to realize that this type of behavior, and the searching image that must exist in the bees' central nervous systems, results from a need never before experienced by these worker bees. The queen may have been through this process before, but worker bees live for only a very few weeks during the warmer months of the year, and swarming may not have occurred for many months or even a few years. Swarming presents the older workers with a totally unprecedented situation, and they proceed to engage in a wholly new type of searching behavior.

An appropriate cavity for a hive is not easily found; bees must search over considerable distances, poking into innumerable crevices in trees, buildings, and rocks before locating anything suitable. In addition to being the right size, a suitable cavity must have only a small entrance near the bottom; it must be dry, and free from ants or other insects. When a worker has found a possible site, she returns to the swarm and communicates to her sisters using the same symbolic dances used to describe the location of food. These dances occur on the surface of the swarm of bees, in many cases on a horizontal part of the swarm where they do not involve the transfer of direction from angle relative to the sun to a vertical angle. But in other respects the dances have the same properties and indicate distance, direction, and desirability in the same manner.

It should be emphasized that the features that make a cavity desirable are vastly different from the concentration of sucrose in nectar or the availability of protein-rich pollen or of water for cooling the colony. To find out how bees evaluate cavities, Seeley (1977) established colonies on small islands where no suitable cavities occurred naturally. He induced swarming by the crude but effective procedure of shaking the queen and numerous workers out of their old hive and leaving them in the open. Various experimental cavities were provided at some distance; the scout bees found these and eventually induced the colony to occupy one of them. Seeley observed that in the preliminary stages the scout bees crawled back and forth over most of the interior of each cavity and spent a considerable amount of time apparently learning what they were like.

Many years previously Lindauer (1955) had experimented with bees having only a limited number of cavities from which to choose, and he compared their behavior when different sorts of cavities were made available. The vigor of the dances and the responses of other bees varied, as one would expect, according to the suitability of the cavity. But the process of evaluation was not completed within a few hours, even though the scout bees had certainly explored the cavities fully in a much shorter time. Instead, over several days various workers visited and danced about the distance, direction, and desirability of several different cavities. Lindauer suggested that this allowed the bees to discover whether the cavities changed in suitability over time, as for example whether a dry cavity might leak on rainy days. This hypothesis has not been tested thoroughly, but the delay allowed Lindauer to discover certain other properties of the communication of swarming honeybees which are of considerable interest from a cognitive viewpoint.

Ordinarily, when several scout bees dance about cavities at different distances and in different directions, and these cavities differ considerably in desirability, the dances describing the better cavities are more vigorous than others. More scouts are recruited to visit the better cavities and to dance when they return. As a result the bees eventually reach a consensus in which virtually all dances are about the best available cavity. After the nearly unanimous dancing has continued for some hours, the whole swarm flies to the best cavity.

From this description one might suppose that the bees returning from less desirable cavities, being mere robots, would simply continue dancing about the cavities they had visited, then gradually cease dancing to produce the observed consensus. But when Lindauer observed several individually marked bees, he found that a dancer sometimes changed from being a communicator to being a recipient. She would stop dancing about her own discovery and follow dances by one of her more enthusiastic sisters. After following several such dances, she would visit the new cavity and begin dancing about it, with the appropriate change in vigor as well as in her indication of distance and direction. Rather than the symbolic gestures being linked rigidly to the stimulus of flowers or cavities, the communicating bee can alternate between the roles of sender and receiver. Even after her initial dancing, her behavior, including her communicative gestures, can be changed by the information she receives through the communication system.

It is important to point out that Lindauer did not observe a bee change her dance pattern until she had actually visited the second cavity. Unlike the weaver ants recruiting their sisters to help repel intruders, these honeybees did not employ chain communication. But Lindauer's experiments involved only a relatively small number of marked bees and a somewhat limited range of conditions. His observations were carried out in the early 1950s, but except for Seeley's studies of how scouts explore cavities, no ethologist has followed up on Lindauer's remarkable discoveries. Therefore it may be premature to conclude that he discovered everything there is to be learned about this fascinating type of behavior. We seem once again to have encountered the inhibiting effect of customary viewpoints even on experimental scientists. For instance, Premack and Woodruff (1978, p. 628 in commentaries) and Premack and Premack (1983) deny that the dances of honeybees should be called a language on the ground that there is no evidence that the bees can judge whether their dances conform to anything in their surroundings. They also doubt that if bees were shown a replica of their own dances they could tell whether such signals accurately represent the distance and direction to a food source they have just visited (Premack and Premack, 1983, pp. 116–122). They do not seem to appreciate the fact that waggle dances are tactile and chemical communication rather than visual signaling, as implied by the stipulation that a bee be shown its own dance pattern. But regardless of this relatively minor distinction, bees dancing about a particular cavity do seem to react differently to their sisters' dances depending on whether such dances are "synonymous" with their own dances. While the question has not yet been studied experimentally, Lindauer did not report that bees returning to a swarm followed the dances of others returning from the same cavity. If it is true that they ignore dances indicating the same direction and distance as their own, this indicates that in some sense they know the meaning of dances like their own. Such speculations could best be tested if an effective model bee could be developed, as I have discussed elsewhere (Griffin, 1981); with such a device we could enter into simple dialogues with bees and question them along the lines suggested by Premack and his colleagues.

The whole subject of bee communication through dances falls so far outside of what we expect of insect behavior that psychologists have virtually ignored it, and even ethologists hold it at arm's length in a rather gingerly fashion. In short, the whole elaborate pattern of symbolic communication employed by honeybees remains a neglected area, despite its implications for additional revolutionary discoveries. Even when von Frisch and his colleagues present incontrovertible evidence that the bees employ a flexible and symbolic communication system, there is enormous resistance to the inference of any conscious thinking. If the same evidence were uncovered in one of our close relatives, such as apes or monkeys, it would be interpreted far differently. We tend to cling tenaciously to the prior convictions that insects are genetically programmed robots, even though this equation of genetic influence with the absence of conscious thinking rests on a very shaky foundation.

One significant reaction to von Frisch's discovery was that of Carl Jung (1973). Late in his life he wrote that although he had believed insects were merely reflex automata:

This view has recently been challenged by the researches of Karl von Frisch . . . bees not only tell their comrades, by means of a peculiar sort of dance, that they have found a feeding-place, but they also indicate its direction and distance, thus enabling beginners to fly to it directly. This kind of message is no different in principle from information conveyed by a human being. In the latter case we would certainly regard such behavior as a conscious and intentional act and can hardly imagine how anyone could prove in a court of law that it had taken place unconsciously . . . We are . . . faced with the fact that the ganglionic system apparently achieves exactly the same result as our cerebral cortex. Nor is there any proof that bees are unconscious.

10 Natural Psychologists

In previous chapters I have mentioned the thought-provoking suggestion by Jolly (1966) and Humphrey (1978) that consciousness may have arisen in human evolution when social groups reached a size and degree of interdependence that made it important for each member to understand his companions' moods, intentions, and thoughts. Humphrey's idea that socially interdependent primates or early men had to be what he calls "natural psychologists" rests on the assumption that for efficient interaction, each group member must be able to understand his companions' frame of mind. He argues that consciousness evolved in our own species because it had adaptive value.

Can we extend this idea to other animals that live in social groups, even to the social insects? They are even more dependent on one another than any vertebrate, except perhaps for our own species. It may be even more important for a worker bee or ant to judge her sisters' moods correctly than it is for primates to assess one another as natural psychologists. In both cases, understanding what another creature is likely to do may best be accomplished by consciously empathizing with it. If so, the small size of the central nervous system in insects may render such economy even more essential than it is for large-brained mammals.

Individual Recognition

One immediate objection to extending the idea of conscious natural psychology to the social insects is the importance attached by Humphrey and others to individual recognition. Primates and other social animals show abundant evidence that they know each other as individuals (Hediger, 1976, 1980; Humphrey, 1974; Cheney and Seyfarth, 1982b). Primate social behavior is strongly conditioned by recognition of each companion's social status and, at least in some monkeys, that of his parents and siblings. This recognition is generally believed to facilitate empathy and assessment of a companion's moods and his likely response to threat, appeasement signals, or requests for food or grooming. A closely linked assumption is that only animals that recognize each other as individuals from one encounter to the next can be effective natural psychologists in Humphrey's sense. Many other animals recognize individuals or at least categories of conspecific animals such as their genetic kin (Beecher, 1982). In a captive colony of fruit-eating bats (*Carollia perspicillata*) Porter (1979) observed that when a harem male heard distress calls from the young offspring of one of the females in his harem, he frequently would crawl to the mother and stimulate her to retrieve her infant.

Social insects are almost universally held to be incapable of individual recognition to any extent. Less social species of insects, on the other hand, do show considerable evidence of individual recognition, especially in the context of mate selection (Lloyd, 1981b). E. O. Wilson states this view categorically on the final page of *The Insect Societies* (1971): "The insect societies are, for the most part, impersonal. The small, relatively primitive colonies of bumblebees and *Polistes* wasps are based on dominance hierarchies, and individuals appear to recognize one another to a limited extent. In other kinds of social insects, however, personalized relationships play little or no role. The sheer size of the colonies and the short life of the members make it inefficient, if not impossible, to establish individual bonds." If this basic tenet of insect sociology is correct, and if individual recognition is essential for animals to function as natural psychologists, then Humphrey's idea cannot be extended to social insects. But perhaps we should examine the basis on which both these assumptions rest.

The colonies of the most advanced social insects include enormous numbers of workers, but careful studies of individually marked worker ants or bees do not seem to have been directed at finding out whether they are completely interchangeable social atoms, all of which will react in precisely the same way. This seems to have been taken for granted, primarily because the colonies are so large and because no one has observed differentiation into subgroupings. But it is not clear how carefully scientists have looked for subgroupings. Workers do specialize to some extent in the tasks they perform and in their social interactions. For example, in the well-studied honeybee, each worker passes through successive occupational stages during her lifetime (reviewed by Lindauer, 1971). The youngest clean cells in the honeycomb: at about three to ten days after emergence they build comb, including capping cells already built, and only at about eighteen days do they begin to spend much time gathering food outside the hive.

These are not rigid sequences; the bees vary what they do according to the needs of the colony. Worker bees of all ages spend a considerable part of their time simply patrolling the hive, feeling and chemically sensing the condition of cells, the cavity walls, and their sisters. They interact with other workers by feeling each other with the antennae and sometimes also through trophallaxis. In these social exchanges the bees do not seem to be identical units; their age and their recent activities subdivide them into at least a few major categories. Younger workers that are busy cleaning cells do not respond to solicitation for trophallaxis in the same way as their older sisters, foragers returning with full stomachs, or those with pollen baskets packed with protein food. Each worker doubtless interacts with many sisters in each major category, but the notion of an infinite pool of completely interchangeable social atoms requires some modification to take account of these differences.

Although no two workers are precisely identical in all particulars, including age, nutritional state, and recent exposure to other social exchanges, there is no evidence that worker A repeatedly interacts with worker B in any way that differs from her behavior toward her sisters that are approximately the same as B in age, nutrition, and recent history. However, such differential reactions would be difficult to detect by any of the customary methods used to study social insects, so we should retain open minds concerning this question, as with so many others regarding the possible thoughts of animals. But to the best of our knowledge, social insects do not get to know one another as primates do.

Let us now turn to the second assumption underlying the belief that a natural psychology in Humphrey's sense is possible only among animals that recognize each other as individuals. Why is individual recognition considered so crucial to consciously understanding the moods of one's companions? Primarily, I suggest, this belief stems from the fact that our fellow men and our close primate relatives display so many idiosyncrasies that to understand their feelings, and perhaps their thoughts, we must pay attention to their individual differences—to what we loosely call personality. Two male monkeys of the same age and social status may react in rather different ways, and these differences remain roughly constant over time. If you, as a vervet monkey, know that young male A tends to be high-strung and aggressive while B is usually moody and less reactive, you can modulate your social intercourse with them effectively—provided you can tell that it really is A or B. In social insects, on the other hand, there is no convincing evidence that a given worker can tell A from B or from a large number of others. Having engaged in trophallaxis with A an hour ago, the ant or bee shows no obvious sign of distinguishing whether the approaching nestmate is A or another sister with which she has never had any contact. But scientists believe so firmly that social insects are undifferentiated social atoms that they have rarely attempted to make close observations of marked individuals comparable to the detailed studies of several species or primates.

A further implicit assumption about social insects is that all those of a given age and recent history react in the same way to a given situation, or that if there is variability it is random and unpredictable. Yet there clearly is considerable variability in insect behavior under ordinary circumstances. Many workers pay no particular attention to each other, and only under special conditions of high motivation or unusually strong stimulation do we observe a highly predictable set of responses. When animals do not respond in any evident manner, the ethologist gathers little or no data, and we usually hear little or nothing about such situations. The result is an easily overlooked bias in our picture of animal behavior.

For example, the recruitment gestures of weaver ants and honeybees described above do not elicit an immediate response from every sister to which they are directed, and under many conditions bees and ants remain motionless or move past each other without any evident interaction. Thus one major difference a social insect encounters in her sisters is the presence or absence of any response at all; nestmates fall into two major categories of responders and nonresponders. Ethologists tend to assume that the nonresponders are of the wrong age or state for appropriate interaction, but it is not always possible to predict from previous observation whether a given insect will respond to the communication signals from another. Social communication between insects is facilitated by rapidly ascertaining whether the other insect is an appropriate partner for a particular social interaction, and many interactions are broken off after a brief contact.

To the extent that workers of a given age and recent history are identical and indistinguishable, their reactions to social signals must also be the same. Therefore any natural psychology need only be concerned with determining the category to which a nestmate belongs; if all those in category P will respond in the same way, while those in category Q will respond in a distinctly different manner, the natural psychologist only has to distinguish P from Q. Perhaps the choice will be made among a larger number of categories, but to the best of our knowledge the number is relatively small. It is as though one's social companions were sets of identical twins or triplets who responded in exactly the same way to one's advances, so that having selected the correct set, it did not matter whether one solicited attention from Tweedledum or Tweedledee.

Assuming that social insects need only classify their sisters into one of several dispositional categories and treat all members of that category alike, it may still be most important for the worker bee or ant to distinguish accurately among the categories. A returning honeybee forager may well avoid antennal contact with young workers that are not yet ready to become foragers because no matter how energetically she solicits them for trophallaxis, or how vigorously she dances to them, they will not respond. In practice the communicator will stop after the earliest stages unless the other insect responds appropriately.

Humphrey suggested that the basic adaptive value of consciousness for early man lay in its usefulness, perhaps even its necessity, for regulating complex social interactions for the mutual benefit of all concerned. In the social insects, these interactions are of even greater importance than in primates or, presumably, in our early ancestors. While the six-legged natural psychologist may need only to distinguish with her millimeter-sized brain between a few categories of companion, rather than between twenty or thirty individually known group members, her ability to do so may be of equal or greater importance. If conscious empathy is helpful in one case, it may be equally helpful, or more so, in the other.

In an entirely different context Premack and Woodruff (1978) argue that chimpanzees impute beliefs and desires to others and that such conscious empathy is more basic than objective assessment of another's behavior. They sum up the argument as follows: "The ape could only be a mentalist. Unless we are badly mistaken, he is not intelligent enough to be a behaviorist" (p. 526).

Thoughtful Porpoises

One reaction to the general ideas I have discussed is that "all those little animals have correspondingly meager thoughts." This is appealing enough until we realize that ours are not the largest brains on this planet. The cetaceans—whales and dolphins, or porpoises—have very large brains; those of the largest whales are several times larger than ours. Both in captivity and under natural conditions cetaceans demonstrate impressive behavioral versatility, and a great deal of it is even more strongly suggestive of conscious thinking than that of beavers, bees, or bowerbirds.

The technical requirements and the financial costs of keeping porpoises in captivity are so formidable that only within the last fifty years has this been done to any extent. The common bottle-nosed dolphin *Tursiops truncatus* has been used extensively in experiments testing its behavioral and mental capabilities, as well as in spectacular performances for public entertainment. Porpoises have been trained to do elaborate and amusing tricks, but sometimes these animals are so perceptive and ingenious that they seem to outwit their human masters. Numerous behavior patterns of cetaceans are difficult to explain without postulating conscious thinking and subjective feeling, but I will describe only a few that seem most suggestive. Many more are discussed in *Cetacean Behavior,* edited by L. M. Herman (1980).

One of the most impressive types of discriminative behavior in cetaceans, though not unique to them, is their ability to imitate another animal. Imitation is a special case of observational learning, but sometimes the copying is so clever that it seems the imitator must think about the process. Herman describes how two ethologists, Tayler and Saayman (1973), observed a bottle-nosed dolphin imitating the behavior of a seal and other animals that shared the same aquarium.

The behaviors imitated included the seal's swimming movements, sleeping posture, and comfort movements (self-grooming). The dolphin, like the seal, at times swam by sculling with its flippers while holding the tail stationary. The sleeping posture imitated was lying on one side at the surface of the water, extending the flippers, and trying to lift the flukes clear of the water. The comfort movement mimicked was vigorous rubbing of the belly with the under surface of one or both flippers . . . Additional imitative behaviors by this dolphin included the swimming characteristics and postures of turtles, skates, and penguins. The dolphin also attempted to remove algae from an underwater window with a sea gull feather, in imitation of the activity of a human diver who regularly cleaned the window. The dolphin, while "cleaning" the window, reportedly produced sounds resembling those from the demand valve of the diver's regulator and emitted a stream of bubbles in apparent imitation of the air expelled by the diver. Tayler and Saayman also observed a dolphin using a piece of broken tile to scrape seaweed from the tank bottom, a behavior apparently derived from observing a diver cleaning the tank with a vacuum hose. The scraping behavior of this dolphin was then copied by a second dolphin.(p. 402)

None of these specialized behaviors have been observed in other members of this or, I suspect, other species of dolphin.

Herman also describes how porpoises often learn to perform some novel and relatively complex pattern of behavior by observing another animal.

In some cases, the animals may even train themselves. In one interesting episode illustrating self-training, one animal of the group had been taught to leap to a suspended ball, grasp it with its teeth, and pull it some distance through the water in order to raise an attached flag. This animal was subsequently removed from the show and a second animal in the group was trained in the same task, but learned to raise the flag by striking repeatedly at the ball with its snout rather than by pulling it. This second animal, a female, subsequently died and another female of the group immediately took over the performance, without training, and continued to strike at the ball with the snout. Later when this new female refused to participate in the show during a two-day period, a young male in the group immediately performed the behavior, but grasped and pulled the ball with his teeth, in the manner of the originally trained animal. (p. 406)

In these cases it is difficult to escape the conclusion that the animals that learned so much by watching their companions and executed the behavior so well at the first opportunity, must have thought consciously about the observed pattern and remembered it in detail.

Karen Pryor's book *Lads before the Wind* (1975) describes her extensive practical work with captive dolphins at a large ocean-arium in Hawaii, where she trained the animals to perform many tricks to entertain visitors. From this extensive experience, she and her colleagues were led to try the sort of experiment that most students of animal behavior would consider too far-fetched to attempt. They trained a captive rough-toothed dolphin (*Steno bredanensis*) named Hou to perform a wide variety of complicated maneuvers in his tank to obtain a reward of food. First, the animal was fed only after performing one particular type of display, such as an aerial backflip, "walking" partly out of water by vigorous coordinated motions of its tail, or slapping its tail while swimming at the surface on its back. Then food was withheld unless the dolphin did something new, something different from any of the tricks he had displayed before. It took Hou some time to realize what was expected of him, but after several weeks he began to invent some new form of aquatic or aerial gymnastics each day to get his food. Hou had evidently formed the concept of newness or "something I've never done before." One other dolphin also made some progress toward producing novel behavior (Pryor, Haag, and O'Reilly, 1969).

Here is an area in which the well-developed methods of experimental psychology could easily be put to use. Could other animals invent novel patterns of behavior rather than execute a particular action, in order to obtain an appropriate reward? Are only porpoises and people capable of such inventiveness? Probably not, because Pryor and her colleagues also trained a pigeon to perform a new act each day. The concept of "something new" may be within the capacities of many animals, but experimental scientists have only begun to design experiments to explore such possibilities.

The pelagic porpoises of the Pacific have become unfortunate victims of newly developed methods of catching tuna. Tuna are predatory fish that hunt in large schools, and for some poorly understood reason they tend to be associated with the Pacific spotted porpoise (*Stenella attenuata*) and the spinner porpoise (*Stenella longirostris*). Recently tuna fishermen have been able to capture most or all of a large tuna school by watching for concentrations of these dolphins, then laying out an enormous net in a circle a few hundred meters in diameter, surrounding both tuna and porpoises, and gradually tightening the net. When the method was first developed, many porpoises were unintentionally killed along with the tuna. After vigorous public objections to this needless slaughter, the more skillful operators of fishing vessels learned to let a part of the net drop so that the porpoises could swim out.

Karen Pryor and several colleagues studied the behavior of porpoises surrounded by a tuna net and saw several indications that they had learned to adapt their behavior to this entirely novel and stressful situation. These porpoises recognize the fishing vessels at hundreds of meters, and when they see one approach, they often change their normal behavior of relatively conspicuous blowing or exhalation as they rise to the surface. Instead they lie just at the surface with only the blow hole above the water and breathe in a way that renders them much less visible to the lookout on a tuna fishing boat. If this stratagem fails, the porpoises swim very rapidly, apparently trying to keep on the right-hand side of the tuna boat. They seem to have learned that the cranes and other machinery used to handle the nets are customarily mounted on the left. If neither of these tactics is successful and the porpoises find themselves surrounded by a net, they act mildly agitated but remain fairly quietly near the center of the net. They no longer dive or swim blindly into the net, as they did when this type of fishing first began. Ordinarily these completely pelagic porpoises never encounter anything other than the moving animals of the open ocean. The ocean bottom, shoreline, and other large solid structures are completely outside their experience, and consequently their behavior toward something like a fishing boat or a net is initially quite inappropriate. But most of the porpoises in the tuna fishing area have learned about the hazard of being surrounded by a net. They wait until the boat "backs down" and allows part of the net to drop below the surface, leaving a small area where they can easily swim out. Most of the tuna remain at deeper levels and do not escape. When the porpoises notice the backing-down maneuver they swim rapidly through the opening. Shortly afterward many of them leap out of the water repeatedly, as though in joyful celebration of their escape. This versatile alteration of their customary behavior is certainly an impressive use of their man-sized brains.

Of the numerous other sorts of cetacean behavior that suggest conscious thought, I will consider only one. This is the occasional provision of assistance to another cetacean by pushing or lifting it to the surface when it is injured or ill, or even when it has died. Although this is hardly an everyday occurrence, aiding behavior has been observed in detail in several groups of captive porpoises and occasionally under natural conditions. It may explain the rare but startling occasions when a human swimmer finds himself lifted from below. Among captive porpoises it has been observed that when one animal becomes visibly weaker and sinks below the surface, others may swim down, push their bodies below that of the animal in trouble, and lift it to the surface. This is done in a sensibly coordinated fashion; they do not push just any part of the animal's body to the surface, but always the dorsal surface where the blow hole is located. Porpoises are also selective and are far more likely to aid females and young animals; adult males are sometimes left on the bottom of the tank. The behavior is sometimes carried to useless extremes, as when a mother carried a stillborn baby for days until decomposition was far advanced. In another instance, a captive bottle-nosed porpoise carried a dead leopard shark for eight days without eating and resisted attempts by human caretakers to remove the dead shark; when they finally succeeded, the porpoise began to eat again. As they devote extraordinary efforts to aid another animal, porpoises must have some thoughts. To be sure, their efforts are often misguided and maladaptive; the stillborn infant could not be revived, and the shark was if anything an enemy. But vivid human thinking occurs at times of extreme grief or triumph and may be accompanied by behavior that accomplishes no practical result.

Porpoises display such great versatility under a wide variety of conditions that they must think out solutions to some of the problems they solve so ingeniously. Experimental scientists continue to explore their mental capacities, limited primarily by the difficulties and expense of working with such large aquatic animals. Despite the lingering constraints of semantic behaviorism, cautious scientists have made impressive progress in long-term studies of captive cetaceans under controlled conditions. A common impediment to research is the greater importance attached to training these animals for public entertainment. Let us hope that increased interest in learning their true capacities will allow more porpoise time to be devoted to carefully controlled experiments.

Can we learn to communicate with porpoises and thus learn about their thoughts and feelings? Pryor (1975) describes how experienced trainers interpret many signs of their charges' emotional states and intentions, but it is difficult to restrain our enthusiasm and distinguish fact from fiction. Extravagant and wholly unconvincing claims have been advanced about porpoises learning to imitate English words and use them with meanings similar to ours. Such claims have given an unfortunate aura of mysticism and sensationalism to efforts to communicate with cetaceans. Despite this handicap, serious efforts are under way to teach captive porpoises to communicate with us in a way that is compatible with their own ways of life, and perhaps their natural modes of thinking. If these efforts succeed, it may be possible to study the thoughts of porpoises.

Wordy Apes

Some of the most convincing recent evidence about animal thinking stems from the pioneering work of Alan and Beatrice Gardner of the University of Nevada (1969, 1979). The Gardners had noted that wild apes seem to communicate by observing each other's behavior, and they suspected that the extremely disappointing results of previous efforts to teach captive chimpanzees to use words reflected not so much a lack of mental ability as a difficulty in controlling the vocal tract. Captive chimpanzees had previously demonstrated the ability to solve complex problems and, like dogs and horses, they had learned to respond appropriately to many spoken words. The Gardners wanted to find out whether apes could also express themselves in ways that we could understand. In the late 1960s they made a concerted effort to teach a young chimpanzee named Washoe to communicate with people using manual gestures derived from American Sign Language. This language, one of many that have been developed in different countries for use by the deaf, consists of a series of gestures or signs, each of which serves the basic function of a single word in spoken or written language. To permit fluent conversation, these signs have evolved into clearly distinguishable hand motions and finger configurations that can be performed rapidly.

Washoe was reared in an environment similar to that in which an American baby would be raised. All the people who cared for Washoe "spoke" to her only in American Sign Language, and used it exclusively when conversing with each other in her presence. They signed to Washoe, much as parents talk to babies who have not yet learned to speak, but always in sign language rather than spoken English. Washoe was encouraged to use signs to ask for what she wanted, and she was helped to do this by a procedure called molding, in which the trainer gently held the chimpanzee's hand in the correct position and moved it to form a certain sign.

The Gardners were far more successful than most scientists would have predicted on the basis of what was previously known about the capabilities of chimpanzees or any other nonhuman species, although Robert Yerkes had anticipated such a possibility (Bourne, 1977). During four years of training Washoe learned to use more than 130 wordlike signs and to recognize these and other signs used by her human companions. She could make the appropriate sign when shown pictures of an object, and on a few occasions she seemed to improvise new signs or new two-sign combinations spontaneously. The best example of this was Washoe's signing "water bird" when she first saw a swan. She also signed to herself when no people were present.

Following the Gardners' lead, several other scientists have trained other great apes to use a quasilinguistic communication system. This work has been thoroughly and critically reviewed by Ristau and Robbins (1982) and widely discussed by many others, so I will give only a brief outline here. Most of the subjects have been female chimpanzees, but two gorillas (Patterson and Linden, 1981) and one orangutan (Miles, 1983) have also been taught gestures based on American Sign Language. Because gestures are variable and require the presence of a human signer, who may influence the ape in other ways that are difficult to evaluate, two groups of laboratory scientists have developed "languages" based on mechanical devices operated by the chimpanzees. David Premack of the University of Pennsylvania used colored plastic tokens arranged in patterns resembling strings of words. His star chimpanzee pupil, named Sarah, learned to select the appropriate plastic "words" to answer correctly when the experimenter presented her with similar chips arranged to form simple questions. Questions such as "What is the color of—?" were answered correctly about familiar objects when the objects were replaced by their plastic symbols, even if the colors were different from those of the objects they represented. Sarah thus learned to answer questions about *represented* objects (reviewed by Premack, 1976; and Premack and Premack, 1983). This type of communication has the property of displacement, as in the case of the honeybee dances.

In another ambitious project at the Yerkes Laboratory of Emory University, Duane Rumbaugh, Sue Savage-Rumbaugh, and their colleagues have used back-lighted keys on a keyboard (Rumbaugh, 1977; Savage-Rumbaugh, Rumbaugh, and Boysen, 1980). Their chimpanzee subjects have learned to press the appropriate keys to communicate simple desires and answer simple questions. In some significant recent studies, two young male chimpanzees, Sherman and Austin, have not only learned to use simple tools to obtain food or toys but have learned to employ the keyboard to ask each other to hand over a certain type of tool. These investigations, as well as extensions of the Gardners' original studies using words derived from American Sign Language, have been extensively reviewed (Ristau and Robbins, 1982) and discussed by Patterson and Linden (1981) and Terrace (1979). Despite disagreement about many aspects of this work, almost everyone concerned agrees that the captive apes have learned, at the very least, to make simple requests and to answer simple questions through these wordlike gestures or mechanical devices.

A heated debate has raged about the extent to which such learned communication resembles human language. Sebeok and Umiker-Sebeok (1980) and Sebeok and Rosenthal (1981) have argued vehemently that the whole business is merely wishful and mistaken reading into the ape's behavior of much more than is really there. They stress that apes are very clever at learning to do what gets them food, praise, social companionship, or other things they want and enjoy. They believe that insufficiently critical scientists have overinterpreted the behavior of their charges and that the apes have really learned only something like: "If I do this she will give me candy," or "If I do that she will play with me," and so forth. They also believe that the apes may be reacting to unintentional signals from the experimenters and that the interpretations have involved what behavioral scientists call "Clever Hans errors." This term refers to a trained horse in the early 1900s that learned to count out answers to arithmetical questions by tapping with his foot. For instance if shown 4×4 written on a slate board, the horse would tap sixteen times. More careful studies showed that Hans could solve such problems only in the presence of a person who knew the answer. The person would inadvertently nod or make other small motions in time with Hans' tapping and would stop when the right number had been reached. Hans had learned to perceive this unintentional communcation, not the arithmetic. The Sebeoks argue that Washoe and her successors have learned, not how to communicate with gestural words, but rather how to watch for signs of approval or disapproval from their human companions and to do what is expected.

Although students of animal behavior must constantly guard against such errors, many of the experiments described above included careful controls that seem to have ruled out this explanation of all of the languagelike communication learned by Washoe and her successors. In many cases the ape's vocabulary was tested by having one person present a series of pictures that the animal was required to name, while a different person, who could not see the pictures, judged what sign Washoe used in response. Furthermore the sheer number of signs that the apes employed correctly would require a far more complex sort of Clever Hans error than an animal's simple noticing that a person has stopped making small-scale counting motions.

Another criticism of the ape language studies has been advanced by Terrace and colleagues (1979). Terrace, aided by numerous assistants, taught a young male chimpanzee named Nim Chimpsky to use about 125 signs over a forty-five-month period. He agrees that Nim, like Washoe and several other language-trained apes, did indeed learn to use these gestures to request objects or actions he wanted and that Nim could use some of them to answer simple questions. But when Terrace analyzed videotapes of Nim exchanging signs with his trainers, he was disappointed to find that many of Nim's "utterances" were copies of what his human companion had just signed. This is scarcely surprising, inasmuch as his trainers had encouraged him to repeat signs throughout his training.

Terrace and his colleagues also concluded that Nim showed no ability to combine more than two signs into meaningful combinations and that his signing never employed even the simplest form of rule-guided sentences. It is not at all clear, however, whether Nim's training provided much encouragement to develop grammatical sentences. In any event, he did not do so, and Terrace doubts whether any of the other signing apes have displayed such a capability. But Miles (1983) reports that her orangutan Chantek's use of gestural signs resembled the speech of young children more closely than Nim's, and Patterson believes that her gorilla Koko follows some rudimentary rules in the sequence of her signs. Yet even on the most liberal interpretation there remains a large gap between the signing of these trained apes and the speech of children who have vocabularies of approximately the same size. The children tend to use longer strings of words, and the third or later words add important meaning to the first two. In contrast, Nim and other language-trained apes seem much more likely to repeat signs or add ones that do not seem, to us at least, to change the basic meaning of a two-sign utterance. For instance, the following is one of the longer utterances reported for the gorilla Koko: "Please milk please me like drink apple bottle"; and from Nim, "Give orange me give eat orange give me eat orange give me you." But grammatical or not, there is no doubt what Koko and Nim were asking for. To quote Descartes and Chomsky (1966), *"The word is the sole sign and certain mark of the presence of thought."* Grammar adds economy, refinement, and scope to human language, but words are basic. Words without grammar are adequate though limited, but there is no grammar without words. And it is clear that Washoe and her successors use the equivalent of words to convey simple thoughts.

The enormous versatility of human language depends not only on large vocabularies of words known to both speakers and listeners but on mutually understood rules for combining them to convey additional meaning. George A. Miller (1967) has used the term "combinatorial productivity" for this extremely powerful attribute of human language. By combining words in particular ways we produce new messages logically and economically. If we had to invent a new word to convey the meaning of each phrase and sentence, the required vocabulary would soon exceed the capacity of even the most proficient human brains. But once a child learns a few words, he can rapidly increase their effectiveness by combining them in new messages in accordance with the language's rules designating which word stands for actor or object, which are modifiers, and so forth.

Signing apes so far have made very little progress in combinatorial productivity, although some of their two-sign combinations seem to conform to simple rules. The natural communication systems of other animals make no use of combinatorial productivity, as far as we know. But the investigation of animal communication has barely begun, especially as a source of evidence about animal thoughts. What has emerged so far has greatly exceeded the prior expectations of scientists; we may be seeing only the tip of yet another iceberg. Extrapolation of scientific discovery is an uncertain business at best, but the momentum of discovery in this area does not seem to be slackening. The apparent lack of any significant combinatorial productivity in the signing of Washoe and her successors might turn out to be a temporary lull in a truly revolutionary development, which began only about fifteen years ago. Perhaps improved methods of investigation and training will lead to more convincing evidence of communicative versatility.

One relevant aspect of all the ape-language studies to date is that the native language of all the investigators has been English, and the signs taught to apes have been derived from American Sign Language. In English, word order is used to indicate actor or object, principal noun or modifying adjective, and many other rule-guided relationships. But this is very atypical; most other human languages rely much more on inflections or modifications of principal words to indicate grammatical relationships. No one seems to have inquired whether signing apes or naturally communicating animals might vary their signals in minor ways to communicate that a particular sign is meant to designate, for instance, the actor rather than the object. This would be a difficult inquiry, because the signals vary for many reasons, and only a laborious analysis of an extensive series of motion pictures or videotapes would disclose whether there were any consistent differences comparable to those conveyed by inflections of words in human speech.

Regardless of these controversies, there seems no doubt that through gestures or manipulation of tokens or keyboards apes can learn to communicate to their human companions a reasonable range of simple thoughts and desires. They also can convey emotional feelings, although an ape does not need elaborate gestures or other forms of symbolic communication to inform a sensitive human companion that it is afraid or hungry. What the artificial signals add to emotional signaling is the possibility of communicating about specific objects and events, even when these are not part of the immediate situation. Furthermore, when Washoe or any other trained ape signs that she wants a certain food, she must be thinking about that food or about its taste or odor. We cannot be certain just what the signing ape is thinking, but the content of her thought must include at least some feature of the object or event designated by the sign she has learned to use. For instance, the Gardners taught Washoe to use a sign that meant flower, to them. But Washoe used it not only for flowers but for pipe tobacco and kitchen fumes. To her it apparently meant smells. Washoe may have been thinking about smells when she used the sign, rather than about the visual properties of colored flowers, but she was certainly thinking about something that overlapped with the properties conveyed by the word *flower* as we use it.

The major significance of the research begun by the Gardners is its confirmation that our closest animal relatives are quite capable of varied thoughts as well as emotions. Many highly significant questions flow from this simple fact. Do apes communicate naturally with the versatility they have demonstrated in the various sorts of languagelike behavior that people have taught them? One approach is to ask whether apes that have learned to use signs more or less as we use single words employ them to communicate with each other. This is being investigated by studying signing apes that have abundant opportunity to interact with each other. Few results have been reported so far, although some signing does seem to be directed to other apes as well as to human companions. When scientists have been looking for something, and when we hear little or nothing about the results, we conclude that nothing important has been discovered. But the lack of results may only mean that chimpanzees can communicate perfectly well without signs. The subject obviously requires further investigation, and we may soon hear about new and interesting developments.

Animal Dreams and Fantasies

The subject of animal thinking as discussed in this book is usually studied by western intellectuals who try hard to be objective and realistic. But the content of much human consciousness does not conform to objective reality. Fear of ghosts and monsters is very basic and widespread in our species. Demons, spirits, miracles, and voices of departed ancestors are real and important to many people, as are religious beliefs that entail faith in the reality and overriding importance of entities far outside of the physical universe studied by objective science. Yet when we speculate about animal thoughts, we usually assume that they would necessarily involve practical down-to-earth matters, such as how to get food or escape predators; we suppose that animal thinking must be a simpler version of our own subjective thinking about the animal's situation.

But there is really no reason to assume that animal thoughts are rigorously realistic. Apes and porpoises often seem playful, mischievous and fickle, and anything but businesslike, practical, and objective. Insofar as animals do think and feel, they may fear imaginary predators, imagine unrealistically delicious foods, or think about objects and events that do not actually occur in the real world around them. The young vervet monkey that gives the eagle alarm call for a harmless songbird may really fear that this flying creature will attack. As we try to imagine the content of animal thoughts, we should consider that they may be less than perfect replicas of reality. Animals may experience fantasies as well as realistic representations of their environment.

This recognition that animal thinking may not be strictly realistic leads to the question of animal dreams. Sleeping dogs sometimes move and vocalize in ways that suggest they are dreaming; their movements resemble those of feeding, running, biting, and even copulation. They sometimes snarl or bark. Some observers of sleeping animals have concluded that these motions and vocalizations accompany dreams related to recent experiences. Human sleepers show two distinct types of EEG potentials. The first, a relatively low-frequency pattern, characterizes deep sleep; the second, called REM sleep, is more irregular and is usually accompanied by rapid eye movements, which can be recorded separately by electrodes located near the eyes. When human subjects are awakened from one of these types of sleep, they are much more likely to report that they were dreaming during REM sleep (Fishbein, 1981; Morrison, 1983). Comparable recordings from sleeping birds and mammals show very similar patterns of REM sleep (Hartman, 1970; Jouvet, 1979; Cohen, 1979), indicating that mammals and birds may dream.

A few rather limited studies of human eye movements during REM sleep suggest that the movements resemble those that would be expected during the activity or experience about which the person is dreaming. For instance, dreaming about watching a tennis match might produce repeated eye movements back and forth from side to side as the dreamer follows the tennis ball. But such experiments have not been developed into a reliable procedure for monitoring the content of human dreams, so dreaming dogs or other animals cannot yet be studied in this way. But if it could be perfected, such monitoring might allow us to determine what the animal was dreaming about. If that could be accomplished, we might be able to study a type of mental experience that exhibits extreme displacement, for nothing in the sleeper's immediate environment need correspond directly to the content of the dream. Stoyva and Kamiya (1968) have proposed that a combined analysis of electrical recording of eye movements and verbal reports of dream content after waking up might eventually lead to objective investigations of human mental experience. Behaviorists have been characteristically uninterested in pursuing this approach. But although we can easily imagine experiments along the lines outlined above that would yield verifiable objective evidence, too few have been carried out to permit any firm conclusions. Perhaps a combination of cognitive ethology and cognitive neurophysiology will eventually fill this gap and provide empirical evidence about the reality and the content of animal dreams. It would indeed be ironic if our strongest evidence that animals think consciously derived from convincing evidence that they dream.

Inspired Innovation

Three examples of especially enterprising innovation by animals are in many ways even more strongly suggestive of conscious thinking than those we have already considered. The first involves birds, and the other two bring us back to the versatile honeybees. In all three cases the animals were confronted by challenges that were not only new to them as individuals, but also quite unprecedented in their evolutionary background.

The German ethologist Otto Koehler and his colleagues at Freiburg in southwestern Germany carried out numerous experiments on the abilities of birds to solve problems that required what he called "wordless thinking" meaning that they thought about objects and relationships but not in terms of words (Koehler, 1956a, 1956b, 1969). In one of the most impressive of these experiments birds were trained to select from a number of covered vessels the one having a certain number of spots on the lid. The spots varied in size, shape, and position, but a well-trained raven could reliably select the pot with any number from one to seven spots. From the results of many such experiments Koehler concluded that these birds had the concept of numbers from two to seven, which he called unnamed numbers. This ability may be comparable in some ways to the very earliest stages of understanding of number in preverbal children (Gelman and Gallistel, 1978). Koehler also felt that animals understand other relatively simple concepts as unnamed thoughts. There is little reason to suppose that thinking about unnamed numbers had been useful enough in the past for natural selection to have favored it strongly. Yet when it became important to think this way to get food, ravens and a few other birds learned to do so. Seibt (1982) has argued that since pigeons can learn as easily to peck three times when shown two lighted spots, and vice versa, there is no basis for the claim that birds have an unnamed number concept in the sense suggested by Koehler. But in both earlier and more recent experiments pigeons have learned to make responses that require some sort of internal representation of number.

H. Hediger (1968, 1976) has reviewed evidence that many mammals can recognize their names when these are used by zoo keepers. Of course, domestic animals and pets very commonly learn to come when called by the names given them by their human masters. Hediger suspected that certain animals had "unnamed names" for other animals and for familiar human companions. But although there is very convincing evidence that many animals recognize others of their species, there is no convincing evidence that one animal addresses another by some individual "name."

This brings up the question of self-awareness, which I touched on briefly in Chapters 1 and 3. Many scientists and scholars who concede that animals can be consciously aware of the world around them and that they can think about objects and events that are not part of their immediate sensory input nevertheless doubt that animals are ever aware of their own awareness. That is, the great majority of animals are held to lack any concept of themselves and any ability to think in such terms as "It is I who am hungry and am looking for a certain type of food." Allowing that an animal can be aware of outside events but denying that it can be self-aware becomes somewhat ridiculous—can the animal be aware of other creatures but not of itself (Griffin, 1981)? As was briefly discussed in Chapter 4, Gallup (1977) has carried out experiments with mirror images, which he interprets as demonstrating that great apes are capable of self-awareness, but that gibbons, monkeys, and the other animals so far tested are not. Some of the chimpanzees who have been trained to use wordlike gestures are reported to have used a sign for themselves. Perhaps language-trained apes could be induced to use such signs to distinguish between their own actions and the same actions carried out by others. Here again, much more extensive data are needed.

We might obtain significant information about self-awareness in animals by devising methods to distinguish between the following two possibilities: (1) an animal feels hungry and is searching consciously for a particular search image that it knows means food, and (2) the animal is consciously aware that it, itself, is hungry and is searching for the pattern that marks the location of food. Could an animal be trained to communicate under appropriate conditions either of two or more messages meaning "I am hungry and searching for food," "You are hungry and searching for food," or "My babies are hungry and searching for food"? To the best of my knowledge such experiments have not been attempted, but if they should yield positive results, this would provide new evidence for self-awareness. Even more significant in many ways would be the discovery that any animal communicates in this manner under natural conditions; this seems unlikely but is worth watching for.

Let us return once again to the well-studied honeybees. J. L. Gould (1979, 1982), whose experiments confirmed von Frisch's conclusion that the waggle dances do convey information from bee to bee, has pointed out that human agriculture often presents honeybees with challenging problems. The anthers of alfalfa flowers spring back vigorously at the visiting insect, thus dusting it with pollen. These flowers are adapted for pollination by larger insects such as bumblebees. When honeybees enter them, they are knocked about so violently that they learn very quickly to avoid alfalfa. But when no other flowers are available, honeybees learn to enter only alfalfa flowers whose anthers have already been tripped by another insect, or when the colony is in extreme need of food, they bite a hole in the back of the alfalfa flowers to reach the nectar. This atypical method is also used in other situations, as reviewed by Inouye (1983).

My final example of thought-provoking animal behavior is one that occurs during experiments on the symbolic communcation of honeybees. To study the waggle dances, experimenters must induce the bees to visit artificial food sources at considerable distances from their hive. This is begun by providing a rich sugar solution in small dishes right at the hive entrance. After the bees have been visiting this dish, it is gradually moved farther and farther from the hive. At first it can be moved only a few centimeters, later a meter or so, without losing the bees. But when it is about 30 meters from the hive, the experimenter can move it by much larger jumps and the same bees return to it after carrying stomachs full of sugar solution to their sisters in the hive. When the feeder is more than 100 or 200 meters from the hive, it can be moved 20 to 30 meters at a time, and many of the bees that have visited it at previous locations begin to search for it beyond where they found it last. They seem to have realized that this splendid new food source moves and that to find it again they should fly farther out from home. Real flowers do not ordinarily leap 20 or 30 meters in a few minutes, so it is difficult to imagine how natural selection would have prepared honeybees to extrapolate the position of a moving food source. Yet it is not absolutely inconceivable that natural selection has played a role. For example, near steep mountain ridges the area of morning sunshine gradually expands as the mountain shadow diminishes. The area where nectar is available as the flowers open may expand with the sunlight. Perhaps the repertoire of genetically programmed foraging tactics encoded in honeybee DNA provides for this special situation. But even if we accept this rather far-fetched explanation, we must still credit the bees with adapting a tactic ordinarily used in the early morning near steep mountains to flat terrain and to other times of day.

The Issue of Falsification

Scientists often insist that any significant hypothesis must be falsifiable. This means that for a hypothesis to be satisfactory, we must be able to anticipate how it might be confirmed or disconfirmed, even if the necessary procedures are not immediately practicable. Scientists rightly lose interest in a theory if no one can suggest how to prove that it is correct or false. It was not entirely unreasonable for Percival Lowell to postulate that there were canals on Mars, given the limited data he had. This was clearly a testable hypothesis, and Lowell or his contemporaries could readily imagine that future astronomers using improved telescopes or spacecraft might determine whether there were canals or not. As we now know, when suitable pictures of the Martian surface became available they disconfirmed Lowell's hypothesis. Some strict behaviorists object to all hypotheses about conscious experiences in animals, and even in people, on the ground that they cannot imagine any procedure by which such hypotheses can be confirmed or falsified. This may tell us something about the limited imaginations of scientists, and outside of narrow scientific circles this argument has been almost as unconvincing as that of the solopsist philosopher who insists that he is the only thinking person in the universe. For we certainly are able to gather useful, if not entirely perfect, information about the thoughts of our human companions.

How then can we hope to disprove or confirm hypotheses about conscious thinking in animals? I have suggested in earlier chapters that versatile coping with new challenges provides suggestive evidence of conscious thinking, but in every case a behaviorist can argue that a completely unconscious organism could behave in the same effectively adaptable fashion. Hence one must fall back on plausibility arguments, such as those that are used in other sciences. For example, it is not possible to confirm or disprove many inferences about evolutionary lines of descent where no specific fossil evidence is available, and the same is true of conjectures about the evolution of our planet and other parts of the universe.

In Chapter 7 I suggested that if precise electrical correlates of conscious thinking in the human brain are eventually defined, the discovery of equivalent electrical signals in animal brains would be convincing evidence of conscious thinking. It is quite possible that such potentials would not be recorded even under the most favorable conditions, and this would of course tend to falsify this hypothesis about animal consciousness.

After considering all the other possibilities, we circle back to animal communication as the most promising window for obtaining information about animal thoughts. Testable hypotheses have already been developed, evaluated, and either confirmed or disproved. The studies of vervet monkey alarm calls by Seyfarth, Cheney, and Marler (1980a, 1980b) described in Chapter 9 are good examples of confirmation. Von Frisch (1967) examined the possibility that honeybees might transmit information about vertical as well as horizontal direction, but failed to confirm this hypothesis. Antimentalists may object that these and comparable data confirm or falsify hypotheses about communication, not about consciousness. But active and specialized communicative behavior, occurring only in the presence of appropriate social companions and often involving exchanges of signals that depend on the prior responses of a companion, strongly suggests that the communication is conscious and intentional. Behavioristic skeptics may reject all such suggestions, but their rejection is comparable to the refusal to recognize any causal role or other significance for conscious thinking in people. This quaint opinion, as Bunge calls it, has long since lost any credibility outside of a narrowing circle of faithful behaviorists. Ethologists and others interested in animals should therefore cease to accept it as a constraining dogma.

Final acceptance of a change in our beliefs about something as important as animal consciousness will require an accumulation of extensive and mutually reinforcing evidence, far beyond anything yet available. But a good start has been made by pioneers like von Frisch, Hölldobler, Seyfarth, Cheney, and Marler. It remains to be seen whether future research will build up a fully credible fabric of data and interpretations supporting hypotheses about animal consciousness, or whether the accumulating evidence will turn out negative, or inconclusive and unconvincing.

Can we come to any general conclusions about the many accomplishments and capacities of animals that I have reviewed and suggested may be accompanied by conscious thinking? One predictable response is "Isn't natural selection marvelous to have forged such capable little computers!" Others will yawn and say "Why yes, of course, we knew this all along about our dog; why are you scientists making such a fuss about the obvious?"

Most of the strongest evidence suggesting animal consciousness stems from their enterprising solutions to newly arisen problems. To be sure, the versatility of any animal has limits, often set by the evolutionary adaptation of their species for a certain range of environmental conditions and the attendant challenges. The human mind also has limits, but these are much broader, and we can often see far beyond the mental horizons of the animals we know about. Learning and memory are crucial to many kinds of problem solving, but genetically programmed or instinctive behavior can also be quite flexible and versatile. A nest-building wasp never normally encounters an experimental ethologist who adds caterpillars to what she has carried to her well-concealed nest, and her behavior is not sufficiently versatile to be adaptive in this special new situation. But within the limits of normally occurring natural situations, many animals, from caddis fly larvae to chimpanzees, can solve newly arisen problems. Honeybees, porpoises, laboratory animals, and probably numerous others can cope with problems that have no precedents in the experience of either the individual or the species.

We can also discern, dimly and just over the horizon, new vistas of experimental analysis that may lead to more direct information about subjective feelings and thoughts. The prospect of using communication as a window on the feelings and thoughts of animals seems the most promising if only because it is so useful with our own companions. In one sense animals may already be using the window, as they succeed in conveying to one another their feelings and simple thoughts. If other animals can get these messages, cognitive ethologists with the advantage of the human brain should be able to do as well.

Bibliography
Index

Bibliography

Adler, J. 1976. Chemotaxis behavior of bacteria. *Sci. Amer.* 234(4):40–47.

Alcock, J. 1969. Observational learning in three species of birds. *Ibis* 111:308–321.

Armstrong, D. M. 1981. *The nature of mind and other essays.* Ithaca, N.Y.: Cornell University Press.

Armstrong, E. A. 1949. Diversionary display. *Ibis* 91:88–97 and 179–188.

Bacrends, G. P. 1941. Fortpflanzungsverhalten und Orientierung der Grabwespe *Ammophila compestris* Jur. *Tijd. voor Entomol.* 84:72–275.

Baker, L. R. 1981. Why computers can't act. *Amer. Philos. Q.* 18:157–163.

Balda, R. P. 1980. Recovery of cached seeds by a captive *Nucifraga caryocatactes.* *Z. Tierpsychol.* 53:331–346.

Beck, B. B. 1980. *Animal tool behavior.* New York: Garland STPM Press.

———— 1982. Chimpocentrism: Bias in cognitive ethology. *J. Hum. Evol.* 11:3–17.

Beecher, M. D., ed. 1982. From individual to species recognition: Theories and mechanisms. *Amer. Zool.* 22(3):475–607.

Bennett, J. 1978. Some remarks about concepts. *Behav. Brain Sci.* 1:557–560.

Bishop, J. 1980. More thought on thought and talk. *Mind* 89:1–16.

Bitterman, M. E. 1965. Phyletic differences in learning. *Amer. Psychol.* 20:396–410.

Boden, M. A. 1972. *Purposive explanations in psychology.* Cambridge, Mass.: Harvard University Press.

————, ed. 1977. *Artificial intelligence and natural man.* New York: Basic Books.

Boring, E. G. 1950. *A history of experimental psychology.* New York: Appleton-Century-Crofts.

Bourne, G. H., ed. 1977. *Progress in ape research.* New York: Academic Press.

Bowman, R. I. 1961. Morphological differentiation and adaptation in the Galapagos finches. *Univ. Calif. Pub. Zool.* 58:1–326.

Bowman, R. I., and S. L. Billeb. 1965. Blood-eating in a Galapagos finch. *Living Bird* 4:29–44.

Bowman, R. S., and N. S. Sutherland. 1970. Shape discrimination by goldfish: Coding of irregularities. *J. Comp. Physiol. Psychol.* 72:90–97.

Bradshaw, J. W. S. 1981. The physiochemical transmission of two components of a multiple chemical signal in the African weaver ant (*Oecophylla longinoda*). *Anim. Behav.* 29:581–585.

Bradshaw, J. W. S., R. Baker, and P. E. Howse. 1975. Multicomponent alarm pheromones of the weaver ant. *Nature* 258:230–231.

—— 1979. Multi-component alarm pheromones in the mandibular glands of major workers of the African weaver ant, *Oecophylla longinoda*. *Physiol. Entomol.* 4:15–25.

Bradshaw, J. W. S., R. Baker, P. E. Howse, and M. D. Higgs. 1979. Caste and colony variations in the chemical composition of the cephalic secretions of the African weaver ant *Oecophylla longinoda*. *Physiol. Entomol.* 4:27–38.

Branch, M. N. 1982. Misrepresenting behaviorism. *Behav. Brain Sci.* 5:372–373.

Bristowe, W. S. 1976. *The world of spiders*. London: Collins.

Bronson, G. W. 1982. The scanning patterns of human infants: Implications for visual learning. Norwood, N.J.: Ablex.

Buchler, E. R. 1976. Prey selection by *Myotis lucifugus* (Chiroptera: Vespertilionidae). *Amer. Nat.* 110:619–628.

Buchwald, J. S., and N. S. Squires. 1982. Endogenous auditory potentials in the cat, A P300 model. In *Conditioning, representation of involved neural functions*, ed. C. D. Woody, pp. 503–515. New York: Plenum.

Bullock, T. H., and G. A. Horridge. 1965. *Structure and function in the nervous systems of invertebrates*. San Francisco: Freeman.

Bunge, M. 1980. *The mind-body problem, a psychological approach*. New York: Pergamon.

Burrows, M. 1977. Flight mechanisms of the locust. In *Identified neurons and behavior of arthropods,* ed. G. Hoyle, pp. 339–356. New York: Plenum.

—— 1982. Interneurones co-ordinating the ventilatory movements of the thoracic spiracles in the locust. *J. Exp. Biol.* 97:385–400.

Callaway, E., P. Tueting, and S. H. Koslow, eds. 1978. *Event-related potentials in man*. New York: Academic Press.

Cerella, J. 1979. Visual classes and natural categories in the pigeon. *J. Exp. Psychol.: Hum. Percep. Perform.* 5:68–77.

—— 1982. Mechanisms of concept formation in the pigeon. In *Analysis of visual behavior,* ed. D. J. Ingle, M. A. Goodale, and R. J. W. Mansfield. Cambridge, Mass.: MIT Press.

Chapman, R. M., J. W. McCrary, J. A. Chapman, and H. R. Bragdon. 1978. Brain responses related to semantic meaning. *Brain Lang.* 5:195–205.

Cheney, D. L., and R. M. Seyfarth. 1982a. How vervet monkeys perceive their grunts: Field playback experiments. *Anim. Behav.* 30:739–751.

—— 1982b. Recognition of individuals within and between groups of free-ranging vervet monkeys. *Anim. Behav.* 32:519–529.

Chisholm, A. H. 1954. The use of "tools" or "instruments." *Ibis* 96:380–383.

—— 1971. Further notes on tool-using by birds. *Victorian Nat.* 88:342–343.

—— 1972. Tool-using by birds: A commentary. *Bird Watcher* 4:156–159.

Chomsky, N. 1966. *Cartesian linguistics*. New York: Harper and Row.

Churchland, P. M. 1979. *Scientific realism and the plasticity of mind*. London: Cambridge University Press.

—— 1983. Matter and consciousness: A contemporary introduction to the philosophy of mind. Bradford Books, MIT Press. In press.

Churchland, P. S., and P. M. Churchland. 1978. Internal states and cognitive theories. *Behav. Brain Sci.* 1(4):565–566.

Clayton, D. A. 1978. Socially facilitated behavior. *Q. Rev. Biol.* 53:373–392.

Cobb, J. B., Jr., and David R. Griffin, eds. 1978. *Mind in nature: Essays on the interface of science and philosophy*. Washington: University Press of America.

Cohen, D. B. 1979. *Sleep and dreaming: Origins, nature, and functions*. New York: Pergamon.

Collias, N., and E. Collias. 1962. An experimental study of the mechanisms of nest building in a weaverbird. *Auk* 79:568–595.

—— 1964. The evolution of nest-building in weaver birds (Ploceidae). *Univ. Calif. Pub. Zool.* 73:1–239.

Corning, W. C., J. A. Dyal, and A. O. D. Willows. 1973–1975. *Invertebrate learning*, vols. 1 (1973), 2 (1973), and 3 (1975). New York: Plenum.

Cowie, R. J., J. R. Krebs, and D. F. Sherry. 1981. Food storing by marsh tits. *Anim. Behav.* 29:1252–1259.

Crane, J. 1975. *Fiddler crabs of the world*. Princeton, N.J.: Princeton University Press.

Crook, J. H. 1964. Field experiments on the nest construction and repair behaviour of certain weaverbirds. *Proc. Zool. Soc. London* 142: 217–255.

Croze, H. 1970. Searching image in carrion crows. *Z. Tierpsychol.* (suppl.) 5:1–86.

Curio, E., U. Ernst, and W. Vieth. 1978a. The adaptive significance of avian mobbing. II. Cultural transmission of enemy recognition in blackbirds: Effectiveness and some contraints. *Z. Tierpsychol.* 48:184–202.

—— 1978b. Cultural transmission of enemy recognition: One function of mobbing. *Science* 202:899–901.

Daanje, A. 1951. On the locomotory movements in birds and the intention movements derived from them. *Behaviour* 3:48–98.

Darwin, C. 1882. *The formation of vegetable mould through the action of worms, with observations of their habits*. New York: Appleton. Reprint 1969, International Publication Service.

Davidson, D. 1975. Thought and talk. In *Mind and language*, ed. S. Guttenplan. London: Oxford University Press.

Davidson, J. M., and R. J. Davidson, eds. 1980. *The psychobiology of consciousness*. New York: Plenum.

Davies, N. B. 1977. Prey selection and social behaviour in wagtails. *J. Anim. Ecol.* 46:37–57.

Delius, J. D., and G. Habers. 1978. Symmetry: Can pigeons conceptualize it? *Behav. Proc.* 1:15–27.

Dennett, D. C. 1978. *Brainstorms, philosophical essays on mind and psychology.* Montgomery, Vt.: Bradford Books.

———— 1983. Intentional systems in cognitive ethology: The "Panglossian Paradigm" defended. *Behav. Brain Sci.,* in press.

Desmedt, J. E. 1981. Scalp-recorded cerebral event-related potentials in man as a point of entry into the analysis of cognitive processing. In *The organization of the cerebral cortex,* ed. F. O. Schmitt et al. Cambridge, Mass.: MIT Press.

Donchin, E. 1981. P300 and classification. In *Electrophysiological approaches to human cognitive processing,* ed. R. Galambos and S. A. Hillyard. *Neurosci. Res. Prog. Bull.* 20:157–161.

Dreyfus, H. L. 1979. *What computers can't do: The limits of artificial intelligence,* rev. ed. New York: Harper and Row.

Duerden, J. E. 1905. On the habits and reactions of crabs bearing actinians in their claws. *Proc. Zool. Soc. London* 2:494–511.

Duncan-Johnson, C. C., and E. Donchin. 1982. The P300 component of the event-related brain potentials as an index of information processing. *Biol. Psychol.* 14:1–52.

Eccles, J. C. 1974. Cerebral activity and consciousness. In *Studies in the philosophy of biology, reduction and related problems,* ed. F. J. Ayala and T. Dobzhansky. Berkeley: University of California Press.

Edwards, C. A., and J. R. Lofty. 1972. *Biology of earthworms.* London: Chapman and Hall.

Elsner, N., and A. V. Popov. 1978. Neuroethology of acoustic communication. In *Advances in insect physiology,* ed. J. E. Treherne, J. J. Berridge, and V. B. Wigglesworth. 13:229–355. New York: Academic Press.

Erber, J. 1975a. The dynamics of learning in the honey bee (*Apis mellifera carnica*). I. The time dependence of choice reaction. *J. Comp. Physiol.* 99:231–242.

———— 1975b. The dynamics of learning in the honey bee (*Apis mellifera carnica*). II. Principles of information processing. *J. Comp. Physiol.* 99:243–255.

Erber, J., T. Masuhr, and R. Menzel. 1980. Localization of short-term memory in the brain of the bee, Apis mellifera. *Physiol. Entomol.* 5:343–348.

Evans, H. E. 1963. *Wasp farm.* Garden City, N.Y.: Natural History Press.

———— 1966. The comparative ethology and evolution of the sand wasps. Cambridge, Mass.: Harvard University Press.

Evans, H. E., and M. J. West Eberhard. 1970. *The wasps.* Ann Arbor, Mich.: University of Michigan Press.

Fellers, J., and G. Fellers. 1976. Tool use in a social insect and its implications for competitive interactions. *Science* 192:70–72.

Fenton, M. B., and J. H. Fullard. 1979. The influence of moth hearing on bat echolocation strategies. *J. Comp. Physiol.* 132:77–86.

Field, T. M., R. Woodson, R. Greenberg, and D. Cohen. 1982. Discrimination and imitation of facial expressions by neonates. *Science* 218:179–181.

Fishbein, W., ed. 1981. *Sleep, dreams, and memory.* New York: SP Medical and Scientific Books.

Fisher, J., and R. A. Hinde. 1949. The opening of milk-bottles by birds. *Brit. Birds* 42:347–357.

Fodor, J. A. 1968. *Psychological explanation, an introduction to the philosophy of psychology*. New York: Random House.

Fraenkel, G. S., and D. L. Gunn. 1940. *The orientation of animals: kineses, taxes and compass reactions*. London: Oxford University Press. Reprint 1961, New York: Dover.

Frisch, K. von. 1967. *The dance language and orientation of bees*. Cambridge, Mass.: Harvard University Press.

—— 1972. *Bees, their vision, chemical senses and language,* 2nd ed. Ithaca, N.Y.: Cornell University Press.

—— 1974. *Animal architecture*. New York: Harcourt Brace.

Galambos, R., and S. A. Hillyard. 1981. Electrophysiological approaches to human cognitive processing. *Neurosci. Res. Prog. Bull.* 20: 141–264, vi.

Gallup, G. G., Jr. 1977. Self-recognition in primates. A comparative approach to the bidirectional properties of consciousness. *Amer. Psychol.* 32:329–338.

Gardner, B. T., and R. A. Gardner. 1979. Two comparative psychologists look at language acquisition. In *Children's language,* ed. K. E. Nelson. New York: Halstead.

Gardner, R. A., and B. T. Gardner. 1969. Teaching sign language to a chimpanzee. *Science* 165:664–672.

Gayou, D. C. 1982. Tool use by green jays. *Wilson Bull.* 94:593–594.

Gelman, R., and C. R. Gallistel. 1978. *The child's understanding of number*. Cambridge, Mass.: Harvard University Press.

Goldman, L. J., and O. W. Henson, Jr. 1977. Prey recognition and selection by the constant frequency bat, *Pteronotus p. parnellii. Behav. Ecol. Sociobiol.* 2:411–419.

Goodall, J. van Lawick. 1968. Behaviour of free-living chimpanzees of the Gombe Stream area. *Anim. Behav. Monogr.* 1:165–311.

—— 1971. *In the shadow of man*. Boston: Houghton Mifflin.

Gould, J. L. 1975. Honey bee communication: the dance-language controversy. *Science* 189:685–693.

—— 1976. The dance-language controversy. *Q. Rev. Biol.* 51:211–244.

—— 1979. Do honeybees know what they are doing? *Nat. Hist.* 88:66–75.

—— 1982. *Ethology, the mechanisms and evolution of behavior*. New York, Norton.

Gould, J. L., and C. G. Gould. 1982. The insect mind: Physics or metaphysics? In *Animal Mind–Human Mind*, ed. D. R. Griffin. New York: Springer-Verlag.

Green, S., and P. Marler. 1979. The analysis of animal communication. In *Handbook of Behavioral Neurobiology*. Vol. 3, *Social behavior and communication*, ed. P. Marler and J. G. Vandenbergh, chap. 3. New York: Plenum.

Grene, M. 1978. Basic concepts for cognitive ethology. *Behav. Brain Sci.* 1:574–575 (see also commentary, 1:611).

Griffin, D. R. 1958. *Listening in the dark*. New Haven, Conn.: Yale University Press. Reprint 1974, New York: Dover.

—— 1978. Prospects for a cognitive ethology. *Behav. Brain Sci.* 1:527–538 (see also commentary, 1:555–629 and 3:615–623).

———— 1981. *The question of animal awareness,* 2nd ed. New York: Rockefeller University Press; paperback, Los Altos, Calif.: William Kaufmann.

————, ed. 1982. *Animal Mind–Human Mind,* introduction. New York: Springer-Verlag.

Gross, C. G., D. B. Bender, and C. E. Rocha-Miranda. 1974. Infero-temporal cortex and vision: a single-unit analysis. In *The neurosciences: Third study program,* ed. F. O. Schmidt and F. G. Worden. Cambridge, Mass.: MIT Press.

Gross, C. G., C. E. Rocha-Miranda, and D. B. Bender. 1972. Visual properties of neurons in the infero-temporal cortex of the macaque. *J. Neurophysiol.* 35:96–111.

Gross, C. G., and M. Mishkin. 1977. The neural basis of stimulus equivalence across retinal translation. In *Lateralization of the nervous system,* ed. S. Harnad, L. Goldstein, J. Jaynes, and G. Krauthamer. New York: Academic Press.

Hain, J. H. W., G. R. Carter, S. D. Kraus, C. A. Mayo, and H. E. Winn. 1982. Feeding behavior of the humpback whale, *Megaptera novaeangliae,* in the western North Atlantic. *Fishery Bull.* 80:259–268.

Haith, M. M., T. Bergman, and M. J. Moore. 1977. Eye contact and face scanning in early infancy. *Science* 198:853–855.

Hansell, M. H. 1968. The house building behaviour of the caddis-fly larva *Silo pallipes* Fabricius. II. Description and analysis of the selection of small particles. *Anim. Behav.* 16:562–577.

———— 1972. Case building behaviour of the caddis-fly larva *Lepidostoma hirtum. J. Zool. London* 167:179–192.

Harnad, S. 1982. Consciousness: An afterthought. *Cog. Brain Theory* 5:29–47.

Hartman, E. 1970. *Sleep and dreaming.* Boston: Little Brown.

Haynes, B. D., and E. Haynes, eds. 1966. *The grizzly bear, portraits from life.* Norman, Okla.: University of Oklahoma Press.

Hediger, H. 1947. Ist das tierliche Bewusstsein unerforschbar? *Behaviour* 1:130–137.

———— 1968. *The psychology of animals in zoos and circuses.* New York: Dover.

———— 1976. Proper names in the animal kingdom. *Experientia* 32:1357–1364.

———— 1980. *Tiere verstehen, Erkenntnisse eines Tierpsychologien.* Munich: Kindler.

Herman, L. M., ed. 1980. *Cetacean behavior, mechanisms and functions.* New York: Wiley.

Hermann, H. R. 1979–1982. *Social insects.* Vol. 1, 1979; 2, 1981; 3, 4, 1982. New York: Academic Press.

Herrnstein, R. J. 1979. Acquisition, generalization, and discrimination reversal of a natural concept. *J. Exp. Psychol.: Anim. Behav. Processes* 5:118–129.

———— 1982. Stimuli and the texture of experience. *Neurosci. and Biobehav. Rev.* 6:105–117.

Herrnstein, R. J., and P. de Villiers. 1980. Fish as a natural category for people and pigeons. In *The psychology of learning and motivation. Advances in research and theory,* vol. 14, ed. G. H. Bowers. New York: Academic Press.

Herrnstein, R. J., D. H. Loveland, and C. Cable. 1976. Natural concepts in pigeons. *J. Exp. Psychol.: Anim. Behav. Processes* 2:285–302.

Hillyard, S. A., and M. Kutas. 1983. Electrophysiology of cognitive processing. *Ann. Rev. Psychol.* 34:33–61.

Hinde, R. A., and J. Fisher. 1951. Further observations on the opening of milk bottles by birds. *Brit. Birds* 44:393–396.

Hölldobler, B., and E. O. Wilson. 1978. The multiple recruitment system of the African weaver ant *Oecophylla longinoda* (Latreille) (Hymenoptera, Formicidae). *Behav. Ecol. Sociobiol.* 3:19–60.

Honig, W. K., and R. K. R. Thompson. 1982. Retrospective and prospective processing in animal working memory. In *The psychology of learning and motivation. Advances in research and theory*. Vol. 16, ed. G. H. Bower. New York: Academic Press.

Hopkins, C. D. 1974. Electric communication of fish. *Amer. Sci.* 62:426–437.

——— 1981. The neuroethology of electric communication. *Trends in NeuroScience* 4(1):4–6.

Hoyle, G., ed. 1977. *Identified neurons and behavior of arthropods*. New York: Plenum.

Huang, I-N., C. A. Koski, and J. R. de Quardo. 1983. Observational learning of a bar-press by rats. *J. Gen. Psychol.* 108:103–111.

Huber, F. 1974. Neural integration (central nervous system). In *The physiology of insecta*, 2nd ed., vol. IV, ed. M. Rockstein, pp. 4–100. New York: Academic Press.

Hulse, S. H., H. Fowler, and W. K. Honig, eds. 1978. *Cognitive processes in animal behavior*. Hillsdale, N.J.: Erlbaum.

Hume, D. 1739. *A treatise of human nature*. Reprinted 1888, with analytical index by L. A. Selby-Bugge. London: Oxford University Press.

Humphrey, N. K. 1974. Species and individuals in the perceptual world of monkeys. *Perception* 3:105–114.

——— 1976. The social function of intellect. In *Growing points in ethology*, ed. P. P. G. Bateson and R. A. Hinde. New York: Cambridge University Press.

——— 1978. Nature's psychologists. *New Scientist* 78:900–903. Reprinted in *Consciousness and the physical world*, ed. B. Josephson and B. S. Ramachandra (1979). New York: Pergamon.

Hutton, R. S., B. M. Wenzel, T. Baker, and M. Homuth. 1974. Two-way avoidance learning in pigeons after olfactory nerve section. *Physiol. Behav.* 13:57–62.

Hyman, L. H. 1940. *The invertebrates: Protozoa through Ctenophora*. New York: McGraw-Hill.

Iersel, J. J. A. van, and J. van dem Assem. 1964. Aspects of orientation in the diggerwasp *Bembix rostrata. Anim. Behav. Suppl.* 1:145–162.

Inouye, D. W. 1983. The ecology of nectar robbing. In *The biology of nectaries*, ed. B. Bentley and T. Elias. New York: Columbia University Press.

Jackson, J. F., and J. H. Jackson. 1978. *Infant culture*. New York: Thomas Y. Crowell.

Janes, S. W. 1976. The apparent use of rocks by a raven in nest defense. *Condor* 78:409.

Jennings, H. S. 1906. *Behavior of lower organisms*. New York: Columbia University Press. Reprint 1962, Bloomington, Ind.: Indiana University Press.

Jolly, A. 1966. Lemur social behavior and primate intelligence. *Science* 153:501–506.

Jones, T., and A. Kamil. 1973. Tool-making and tool-using in the northern blue jay. *Science* 180:1076–1078.

Jouvet, M. 1979. What does a cat dream about? *Trends in NeuroScience* 2:280–282.

Jung, C. G. 1973. *Synchronicity, a causal connecting principle.* Princeton, N.J.: Princeton University Press.

Jurasz, C. M., and V. P. Jurasz. 1979. Feeding modes of the humpback whale, *Megaptera novaeangliae* in southeast Alaska. *Sci. Repts., Whales Research Institute* 31:69–83.

Kandel, E. 1979a. Small systems of neurons. *Sci. Amer.* 241(3):66–76.

———— 1979b. *Behavioral biology of Aplesia, a contribution to the comparative study of opistobranch molluscs.* San Francisco: Freeman.

Kawai, M. 1965. Newly acquired pre-cultural behavior of the natural troop of Japanese monkeys on Koshima Islet. *Primates* 6(1):1–30.

Kenyon, K. W. 1969. *The sea otter in the eastern Pacific Ocean.* North American Fauna, no. 68. Washington, D.C.: U.S. Bureau of Sport Fisheries and Wildlife.

Kety, S. S. 1960. A biologist examines the mind and behavior. *Science* 132:1861–1870.

Klemm, W. R. 1969. *Animal electroencephalography.* New York: Academic Press.

Klosterhalfen, S., W. Fischer, and M. E. Bitterman. 1978, Modification of attention in honey bees. *Science* 201:1241–1243.

Koehler, O. 1956a. Sprache und unbenanntes Denken. In *L'instinct dans le comportement des animaux et de l'homme,* pp. 647–674. Paris: Masson.

———— 1956b. Thinking without words. *Proc. 14th Int. Zool. Cong., Copenhagen, 1953.* pp. 75–88.

———— 1969. Tiersprachen und Menschensprachen. In *Kreatur Mensch, moderne Wissenchaft auf der Suche nach dem Humanum,* pp. 119–133 and 187–190. Munich: Heinz Moos.

Knudsen, E. I., and M. Konishi. 1978. Space and frequency are represented separately in auditory midbrain of the owl. *J. Neurophysiol.* 41:870–884.

Konishi, M. 1973. How the owl tracks its prey. *Amer. Sci.* 61:414–424.

Krebs, J. R. 1977. Review of *The question of animal awareness. Nature* 266:792.

———— 1978. Optimal foraging: Decision rules for predators. In *Behavioural ecology, an evolutionary approach,* ed. J. R. Krebs and N. B. Davies. Oxford: Blackwell.

———— 1979. Foraging strategies and their social significance. In *Handbook of behavioral neurobiology.* Vol. 3, *Social behavior and communication,* ed. P. Marler and J. G. Vandenbergh. New York: Plenum.

Krebs, J. R., and N. B. Davies. 1978. *Behavioural ecology, an evolutionary approach.* Oxford: Blackwell.

Krebs, J. R., M. H. MacRoberts, and J. M. Cullen. 1972. Flocking and feeding in the great tit (*Parus major*)—an experimental study. *Ibis* 114:507–530.

Kroodsma, D. 1976. Reproductive development in a female songbird: Differential stimulation by quality of male song. *Science* 192:574–575.

Kroodsma, D. E., and E. H. Miller, eds. 1983. *Acoustic communication in birds,* vols. 1 and 2. New York: Academic Press.

Kruuk, H. 1972. *The spotted hyena: a study of predation and social behavior.* Chicago: University of Chicago Press.

Kuhl, P. K., and A. N. Meltzoff. 1982. The bimodal perception of speech. *Science* 218:1138–1141.

Kutas, M., and S. A. Hillyard. 1980. Reading senseless sentences: Brain potentials reflect semantic incongruity. *Science* 207:203–205.

Lack, D. 1947. *Darwin's finches.* New York: Cambridge University Press.

Lindauer, M. 1955. Schwarmbienen auf Wohnungssuch. *Z. vergl. Physiol.* 37:263–324.

———— 1971. *Communication among social bees,* 2nd ed. Cambridge, Mass.: Harvard University Press.

———— 1974. Social behavior and mutual communication. In *The physiology of insecta,* 2nd ed., vol. IV, pp. 149–228. New York: Academic Press.

Lloyd, J. E. 1980. Insect behavioral ecology: Coming of age in bionomics, or compleat biologists have revolutions too. *Florida Entomol.* 63:1–4.

———— 1981a. Firefly mate-rivals mimic their predators and vice versa. *Nature* 290:498–500.

———— 1981b. Sexual selection: Individuality, identification, and recognition in a bumblebee and other insects. *Florida Entomol.* 64:89–118.

———— 1983. Bioluminescence and communication in insects. *Ann. Rev. Entomol.* 28:131–160.

Loeb, J. 1918. *Forced movements, tropisms, and animal conduct.* Philadelphia: Lippincott. Reprint 1973, New York: Dover.

Lorenz, K. 1963. Haben tiere ein subjectives Erleben? *Jahr. Techn. Hochs. München.* Eng. trans., Do animals undergo subjective experience? In *Studies in animal and human behavior, vol. II. Cambridge, Mass.: Harvard University Press.*

———— *1971. Studies in animal and human behavior,* vol. II. Cambridge, Mass.: Harvard University Press.

Lovell, H. B. 1958. Baiting of fish by a green heron. *Wilson Bull.* 70:280–281.

MacKenzie, B. D. 1977. *Behaviorism and the limits of scientific method.* London: Routledge and Kegan Paul.

Mackintosh, N. J. 1974. *The psychology of animal learning.* New York: Academic Press.

Maier, N. R. F., and T. C. Schneirla. 1935. *Principles of animal psychology.* New York: McGraw-Hill. Reprint with supplement, 1964, New York: Dover.

Markl, H. 1974. Insect behavior: Functions and mechanisms. In *The physiology of insecta,* 2nd ed., vol. 4, pp. 3–148. New York: Academic Press.

Marler, P. 1970. Bird song and speech development: Could there be parallels? *Amer. Sci.* 58:669–673.

———— 1976. Sensory templates in species-specific behavior. In *Simpler networks and behavior,* ed. J. C. Fentress, pp. 314–329. Sunderland, Mass.: Sinauer.

———— 1977. The evolution of communication. In *How animals communicate,* ed. T. A. Sebeok. Bloomington, Ind.: Indiana University Press.

———— 1978. Perception and innate knowledge. In *The nature of life,* ed. W. H. Heidcamp. Baltimore: University Park Press.

Marler, P., and S. Peters. 1981. Sparrows learn adult song and more from memory. *Science* 146:1483–1486.

Marshall, A. A. 1954. *Bower birds*. London: Oxford University Press.

Mason, W. A. 1976. Review of *The question of animal awareness. Science* 194:930–931.

Mason, W. A., and R. F. Reidinger. 1982. Observational learning of food aversions in red-winged blackbirds (*Agelaius phoeniceus*). *Auk* 99:548–554.

McMahan, E. A. 1982. Bait-and-capture strategy of termite-eating assassin bug. *Insectes Sociaux* 29:346–351.

——— 1983. Bugs angle for termites. *Nat. Hist.* 92(5):40–47.

Mehler, J., ed. 1983. *Infant perception and cognition*. Hillsdale, N.J.: Erlbaum, in press.

Meltzoff, A. N., and M. K. Moore. 1977. Evidence of perception and thinking in very young human infants. *Science* 198:75–78.

Menzel, R. 1979. Behavioral access to short-term memory in bees. *Nature* 281:368–369.

Menzel, R., J. Erber, and T. Masuhr. 1974. Learning and memory in the honey bee. In *Experimental analysis of insect behavior*, ed. L. Barton Browne. New York: Springer.

Miles, H. L. 1983. Apes and language: The search for communicative competence. In *Language in primates: Implications for linguistics, anthropology, psychology, and philosophy*, ed. J. de Luce and H. T. Wilder. New York: Springer.

Miller, G. A. 1967. *The psychology of communication*. New York: Basic Books.

Miller, L. A., and J. Oleson. 1979. Avoidance behavior in green lacewings. I. Behavior of free flying green lacewings to hunting bats and ultrasound. *J. Comp. Physiol.* 131:113–120.

Millikan, G. C., and R. I. Bowman. 1967. Observations on Galapagos tool-using finches in captivity. *Living Bird* 6:23–41.

Mills, E. A. 1919. *The grizzly, our greatest wild animal*. Boston: Houghton Mifflin.

Morrison, A. R. A window on the sleeping brain. *Sci. Amer.* 248(4):94–102.

Mountcastle, V. B. 1981. Comments in Galambos and Hillyard. *Neurosci. Res. Prog. Bull.* 20:164, 176.

Nagel, T. 1974. What is it like to be a bat? *Philos. Rev.* 83:435–450.

Netzel, H. 1977. Die Bildung des Gehäuses bei *Difflugia oviformes* (Rhizopoda, Testacea). *Arch. Protistenk.* 119:1–30.

Neville, H. J., and S. L. Foote. In press. Auditory event related potentials in the squirrel monkey: Parallels to the human late wave responses. *Brain Res.*

Norman, D. A., ed. 1981. *Perspectives on cognitive science*. Hillsdale, N.J.: Erlbaum.

Norman, J. R. 1949. *A history of fishes*. New York: A. A. Wyn.

Norton-Griffiths, M. 1967. Some ecological aspects of the feeding behaviour of the oyster-catcher *Haematopus ostralegus* on the edible mussel *Mytilus edulis. Ibis* 109:412–424.

——— 1969. The organization, control and development of parental feeding in the oystercatcher (*Haematopus ostralegus*). *Behaviour* 34:55–114.

Orians, G. H. 1980. *Some adaptations of marsh-nesting blackbirds*. Princeton, N.J.: Princeton University Press.

Ornstein, R., J. Johnstone, J. Herron, and C. Swencionis. 1980. Differential right hemisphere engagement in visiospatial tasks. *Neuropsychologia* 18:49–64.

O'Shea, M., and H. F. Rowell. 1977. Complex neural integration and identified interneurons in the locust brain. In *Identified neurons and behavior of arthropods*, ed. G. Hoyle, pp. 307–328. New York: Plenum.

Owings, D. H., and D. W. Leger. 1980. Chatter vocalizations of California ground squirrels: predator- and social-role specificity. *Z. Tierpsychol*. 54:163–184.

Patterson, F. G., and E. Linden. 1981. *The education of Koko*. New York: Holt, Rinehart and Winston.

Payne, R. S., ed. 1983. *Communication and behavior of whales*. Boulder, Colo.: Westview Press.

Pearson, K. G. 1977. Interneurons in the ventral nerve cord of insects. In *Identified neurons and behavior of arthropods*, ed. G. Hoyle, pp. 329–337. New York: Plenum.

Pepperberg, I. M. 1981. Functional vocalizations by an African Grey parrot (*Psittacus erithacus*). *Z. Tierpsychol*. 55:139–160.

———— 1983. Cognition in the African Grey parrot: preliminary evidence for (auditory/vocal) comprehension of the class concept. *Anim. Learn. Behav*.

Picton, T. W., and S. A. Hillyard. 1974. Human auditory evoked potentials. II. Effects of attention. *Electroenceph. Clin. Neurophysiol*. 36:191–199.

Poole, J., and D. G. Lander. 1971. The pigeon's concept of pigeon. *Psychonomic Science* 25:157–158.

Popper, K. R. 1972. *Objective knowledge*. London: Oxford University Press.

———— 1974. Scientific reduction and the essential incompleteness of all science. In *Studies in the philosophy of biology, reduction and related problems*, ed. F. J. Ayala and T. G. Dobzhansky, chap. 16. Berkeley: University of California Press.

Popper, K. R., and J. C. Eccles. 1977. *The self and its brain*. Hillsdale, N.J.: Erlbaum.

Porter, F. L. 1979. Social behavior in the leaf-nosed bat *Carollia perspicillata*. II. Social communication. *Z. Tierpsychol*. 50:1–8.

Premack, D. 1976. *Intelligence in ape and man*. Hillsdale, N.J.: Erlbaum.

Premack, D., and A. J. Premack. 1983. *The mind of an ape*. New York: Norton.

Premack, D., and G. Woodruff. 1978. Does the chimpanzee have a theory of mind? *Behav. Brain Sci*. 1:515–526 (see also commentary, 1:555–629 and 3:615–623).

Pribram, K. H. 1978. Consciousness, classified and declassified. *Behav. Brain. Sci*. 1:590–592.

Pryor, K. 1975. *Lads before the wind*. New York: Harper and Row.

Pryor, K., R. Haag, and J. O'Reilly. 1969. The creative porpoise: Training for novel behavior. *J. Exp. Anal. Behav*. 12:653–661.

Pyke, G. H. 1979. Optimal foraging in bumblebees: Rule of movement between flowers within inflorescences. *Anim. Behav*. 27:1167–1181.

Rachlin, H. 1978. Who cares if the chimpanzee has a theory of mind? *Behav. Brain Sci.* 1:593–594.

Rice, W. R. 1982. Acoustical location of prey by the marsh hawk: Adaptation to concealed prey. *Auk* 99:403–413.

Ristau, C. A., and D. Robbins. 1982. Language in the great apes: A critical review. *Advances in Study of Behavior* 12:142–225.

Roberts, G. J. 1982. Apparent baiting by a black kite. *Emu* 82:53–54.

Rockstroh, B., T. Elbert, N. Birbaumer, and W. Lutzenberger. 1982. *Slow brain potentials and behavior.* Baltimore: Urban and Schwarzenberg.

Roeder, K. D. 1967. *Nerve cells and insect behavior.* Cambridge, Mass.: Harvard University Press.

Roitblat, H. L. 1982. The meaning of representation in animal memory. (With commentaries and author's responses.) *Behav. Brain Sci.* 5:353–406.

Roitblat, H. L., T. G. Bever, H. S. Terrace. 1983. *Animal cognition.* Hillsdale, N.J.: Erlbaum.

Romanes, G. J. 1884. *Mental evolution in animals.* Reprint 1969, with posthumous essay on instinct by Charles Darwin. New York: ABS Press.

Rose, S., P. A. Katz, M. Birke, and E. Rossman. 1977. Visual following in newborns: Role of figure-ground contrast and configurational detail. *Percept. Motor Skills* 45:515–522.

Ross, D. 1971. Protection of hermit crabs (*Dardanus* spp.) from octopus by commensal sea anemones (*Calliactus* spp.). *Nature* 230:401–402.

Rue, L. L. 1964. *The world of the beaver.* Philadelphia: Lippincott.

Rumbaugh, D. M. 1977. *Language learning by a chimpanzee: The Lana Project.* New York: Academic Press.

Sales, G., and D. Pye. 1974. *Ultrasonic communication by animals.* London: Chapman and Hall.

Savage-Rumbaugh, E. S., D. M. Rumbaugh, and S. Boysen. 1978. Linguistically mediated tool use and exchange by chimpanzees (*Pan troglodytes*). *Behav. Brain Sci.* 1:539–554 (see also commentary, 1:555–629 and 3:615–623).

———— 1980. Do apes use language? *Amer. Sci.* 68:49–61.

Savory, T. H. 1959. *Instinctive living, a study of invertebrate behaviour.* London: Pergamon.

Schaller, G. B. 1972. *The Serengeti lion: A study of predator-prey relations.* Chicago: University of Chicago Press.

Schöne, H. 1980. *Orientierung im Raum, Formen und Mechanismen der Lenkung des Verhaltens im Raum bei Tier und Mensch.* Stuttgart: Wissenschaftliche Verlagsgesellschaft.

Schwartz, B., and H. Lacey. 1982. *Behaviorism, science and human nature.* New York: Norton.

Sebeok, T. A., ed. (1977). *How animals communicate.* Bloomington, Ind.: Indiana University Press.

Sebeok. T. A., and R. Rosenthal. 1981. The Clever Hans phenomenon: Communication with horses, whales, apes, and people. *Ann. N.Y. Acad. Sci.* 364:1–311.

Sebeok, T. A., and J. Umiker-Sebeok, eds. 1980. *Speaking of apes, a critical anthology of two-way communication with man.* New York: Plenum.

Seeley, T. 1977. Measurement of nest cavity volume by the honey bee (*Apis mellifera*). *Behav. Ecol. Sociobiol.* 2:201–227.

Seibt, U. 1982. Zahlbegriff und Zählverhalten bei Tieren. Neue Versuche und Deutungen. *Z. Tierpsychol.* 60:325–341.

Seyfarth, R. M., D. L. Cheney, and P. Marler. 1980a. Monkey responses to three different alarm calls: Evidence for predator classification and semantic communication. *Science* 210:801–803.

——— 1980b. Vervet monkey alarm calls: Semantic communication in a free-ranging primate. *Anim. Behav.* 28:1070–1094.

Shallice, T. 1972. Dual functions of consciousness. *Psychol. Rev.* 79:383–393.

Sherry, D. F., J. R. Krebs, and R. J. Cowie. 1981. Memory for the location of stored food in marsh tits. *Anim. Behav.* 29:1260–1266.

Shettleworth, S. J. 1983. Memory in food-hoarding in birds. *Sci. Amer.* 248(3):102–110.

Shorter, J. M. 1967. Other minds. In *The encyclopedia of philosophy*, ed. P. Edwards, vol. 6, pp. 7–13. New York: Macmillan.

Skinner, B. F. 1957. *Verbal behavior*. New York: Appleton-Century-Crofts.

——— 1966. The phylogeny and ontogeny of behavior. *Science* 153:1205–1213.

——— 1974. *About behaviorism*. New York: Random House.

——— 1981. Selection by consequences. *Science* 213:501–504.

Skutch, A. F. 1976. *Parent birds and their young*. Austin, Tex.: University of Texas Press.

Smith, W. J. 1977. *The behavior of communicating: An ethological approach*. Cambridge, Mass.: Harvard University Press.

Sordahl, T. A. 1980. Antipredator behavior and parental care in the American avocet and black-necked stilt (Aves: Recurvirostridae). Ph.D. thesis, Utah State University, Logan, Utah.

——— 1981. Sleight of wing. *Nat. Hist.* 90:42–49.

Sperry, R. 1983. *Science and moral priority, merging mind, brain, and human values*. New York: Columbia University Press.

Stoyva, J., and J. Kamiya. 1968. Electrophysiological studies of dreaming as the prototype of a new strategy in the study of consciousness. *Psychol. Rev.* 75:192–205.

Straub, R. O., and H. S. Terrace. 1981. Generalization of serial learning in the pigeon. *Anim. Learn. Behav.* 9:454–468.

Struhsaker, T. T. 1967. *The red colobus monkey*. Chicago: University of Chicago Press.

Suarez, S. D., and G. G. Gallup, Jr. 1981. Self-recognition in chimpanzees and orangutans, but not gorillas. *J. Hum. Evol.* 10:175–188.

Surlykke, A., and L. A. Miller. 1982. Central branching of three sensory axons from a moth ear (*Agrotis segetum,* Noctuidae). *J. Insect Physiol.* 28:357–364.

Sutton, S., M. Braren, J. Zubin, and E. R. John. 1965. Evoked-potential correlates of stimulus uncertainty. *Science* 150:1187–1188.

Tayler, C. K., and G. S. Saayman. 1973. Imitative behavior of Indian Ocean bottlenose dolphins (*Tursiops aduncus*) in captivity. *Behaviour* 44:286–297.

Terrace, H. S. 1979. *Nim*. New York: Knopf.

Terrace, H. S., L. A. Petitto, and T. G. Bever. 1979. Can an ape create a sentence? *Science* 206:891–902.

Thatcher, R. W., and E. R. John. 1977. *Foundations of cognitive processes*. Hillsdale, N.J.: Erlbaum.

Thorpe, W. H. 1963. *Learning and instinct in animals*. Cambridge, Mass.: Harvard University Press.

Tinkelpaugh, O. L. 1928. An experimental study of representative factors in monkeys. *J. Comp. Psychol.* 8:197–236.

Tolman, E. C. 1932. *Purposive behavior in animals and men*. New York: Appleton-Century.

―――― 1937. The acquisition of string-pulling by rats—conditioned reflex of sign gestalt. *Psychol. Rev.* 44:195–211.

―――― 1959. Principles of purposive behavior. In *Psychology: A study of a science. Study I. Conceptual and systematic*. Vol. 2, *General systematic formulations, learning, and special processes*, ed. S. Koch. New York: McGraw Hill.

―――― 1966. *Behavior and psychological man*. Berkeley: University of California Press.

Treherne, J. E., ed. 1974. *Insect neurobiology*. Amsterdam: North Holland.

Underwood, G., and R. Stevens, eds. 1979. *Aspects of consciousness*. Vol. 1, *Psychological issues*. New York: Academic Press.

Uttal, W. R. 1978. *The psychology of mind*. New York: Wiley.

Van der Kloot, W. G., and C. M. Williams. 1953. Cocoon construction by the *Cecropia* silkworm. *Behaviour* 5:141–174.

Vander Wall, S. B. 1982. An experimental analysis of cache recovery in Clark's nutcracker. *Anim. Behav.* 30:84–94.

Vander Wall, S. B., and R. P. Balda. 1981. Ecology and evolution of food-storing behavior in conifer-seed-caching corvids. *Z. Tierpsychol.* 56:217–242.

Van Lawick-Goodall, J. 1970. Tool-using in primates and other vertebrates. In *Advances in the Study of Behavior*, vol. 3, ed. D. Lehrman, R. Hinde, and E. Shaw, pp. 195–249. New York: Academic Press.

Waal, F. de. 1982. *Chimpanzee politics, power and sex among the apes*. New York: Harper and Row.

Walker, S. 1983. *Animal thought*. London: Routledge and Kegan Paul.

Walther, F. R. 1969. Flight behaviour and avoidance of predators in Thompson's gazelle (*Gazella thompsoni* Guenther 1884). *Behaviour* 34:184–221.

Wasserman, E. A. 1981. Comparative psychology returns: a review of Hulse, Fowler, and Honig's *Cognitive processes in animal behavior*. *J. Exp. Anal. Behav.* 35:243–257.

―――― 1983. Is cognitive psychology behavioral? *Psychol. Record*. 33:6–11.

Weber, N. A. 1972. Gardening ants, the attines. *Memoires Amer. Philos. Soc.* 92:1–146.

Wheeler, W. M. 1910. *Ants, their structure, development, and behavior*. New York: Columbia University Press.

―――― 1928. *The social insects: their origin and evolution*. London: Kegan Paul, Trench Trubner.

Whitehead, A. N. 1938. *Modes of thought*. New York: Macmillan.

Whiteley, C. H. 1973. *Mind in action, an essay in philosophical psychology*. London: Oxford University Press.

Wiggins, G. B. 1977. *Larvae of North American caddis-flies*. Toronto: University of Toronto Press.

Wilder, M. B., G. R. Farley, and A. Starr. 1981. Endogenous late positive component of the evoked potential in cats corresponding to the P300 in humans. *Science* 211:605–606.

Wiley, R. H. 1974. Evolution of social organization and life-history patterns among grouse. *Q. Rev. Biol.* 49:201–227.

Wilson, E. O. 1971. *The insect societies*. Cambridge, Mass.: Harvard University Press.

Wilsson, L. 1968. *My beaver colony*. Garden City, N.Y.: Doubleday.

———— 1971. Observations and experiments on the ethology of the European beaver (*Castor fiber* L.), a study in the development of phylogenetically adapted behaviour in a highly specialized mammal. *Viltrevy, Swedish Wildlife* 8:117–266. (Uppsala, Svenska Jägareförbundet).

Wittenberger, J. F. 1981. *Animal social behavior*. Boston: Duxbury Press.

Woodfield, A. 1976. *Teleology*. London: Cambridge University Press.

Wright, W. H., and J. B. Kenfoot. 1909. *The grizzly bear*. New York: Scribner's.

Zach, R. 1978. Selection and dropping of whelks by northwestern crows. *Behaviour* 67:134–147.

Zentall, T. R., D. E. Hogan, C. A. Edwards, and E. Hearst. 1980. Oddity learning in the pigeon as a function of the number of incorrect alternatives. *J. Exp. Psychol.: Anim. Behav. Proc.* 6:278–299.

Index

Adaptability, as criterion of consciousness, 37
Adaptiveness of behavior, 24
Adler, J., 49
Aesthetics, in bowerbird bowers, 128
Aggressive displays of courting males, 158–159
Alarm calls of vervet monkeys, 166–168
Alcock, J., 138
Alfalfa, exploitation by honeybees, 206
Alpha rhythms, 145
Ambiguity in signs used by apes, 201
American Sign Language, taught to apes, 196–202
Ammophila campestris, 101
Amoebae, protective shells, 97
Anecdotal evidence, 14
Animal cognition, 134–143
Animal communication, as evidence of thinking, 3, 4, 38, 154–164, 208
Animal learning, 134–143, 209
Animals as mechanisms, 9
Anthropocentric thinking, 43
Anticipation, indications of, 37–38
Antipredator behavior: birds, 87–94; porpoises, 194
Ant-lions, tool use by, 119
Ants: leaf-cutter, 103–105; weaver, 106–107, 170–172; agriculture of, 104–105; central nervous system, 105; tool use

by, 120; recruitment by, 170–172; tactile communication, 176; tandem running, 176
Apes: languagelike behavior, 196–202; self-recognition, 205–206
Aphaenogaster, tool use by, 120
Aphids, "domesticated" by ants, 105
Armstrong, D. M., 3–8
Armstrong, E. A., 91
Artifacts of animals, 95–96
Assassin bugs, fishing for termites, 123
Assem, J. van dem, 102
Atta, 103–105
Attention: EEG correlates of, 145
Auditory sensitivity: owls, 75; moths, 76
Automatic behavior, 36–42
Avoidance conditioning: ERP correlates of, 150
Awareness: animal, 1–3; of instincts, 42–43

Baerends, G. P., 101, 102, 112
Bait, use of, 123–124
Baker, L. R., 11
Baker, R., 175
Balda, R. P., 69–71
Bats: minds, 2; discriminative echolocation, 76–78; insect-catching tactics, 78; individual recognition, 187

Beavers: engineering, 128–132; predation on, 128–129; food stores, 129; responses to sounds, 130–131
Beck, B. B., 60, 118, 124, 125
Bee bread, 102
Beecher, M. D., 187
Bees, mason, nests of, 102–103. *See also* Honeybees
Behavior: verbal, 19; as indication of consciousness, 32–40; adaptability of, 35; automatic, 36; genetics of, 174–178, 207
Behavioral ecologists, 24–26, 52–58
Behavioral evolution, 26, 51
Behavioral genetics, 51, 93, 105
Behaviorism: 4–24; inclusive, 21, 38, 93; purposive, 19, 42; as straitjacket, 32; semantic, 134
Bender, D. B., 144
Bennett, J., 3
Bergman, T., 2
Bever, T. G., 3, 4, 134, 199
Billeb, S. L., 63–64
Biological clocks, 137
Birds: foraging behavior of, 53; nests of, 107–110. *See also entries under common names*
Bird song: learning and genetics, 114; as communicative display, 156–157; complexity of and female choice, 160
Bishop, J., 2
Bitterman, M. E., 6, 7
Blackbirds: feeding, 54–58; nesting behavior, 55; territory evaluation, 55–58; antipredator behavior, 87; mistakes in nest building, 109; female choice, 157
Blue jays, tool use by, 122
Boden, M. A., 11, 19
Boobies, blood taken by finches, 63–65
Boring, E. G., 143
Bottles, used by sea otters, 125
Bourne, G. H., 197
Bowerbirds, bowers of, 127–128
Bowman, R. I., 63–64, 122
Bowman, R. S., 141
Boysen, S., 3, 198
Bradshaw, J. W. S., 175
Bragdon, N. R., 148–152
Brain mechanisms of consciousness, 2, 23, 147–153
Brain waves, 144–145
Branch, M. N., 22
Bristowe, W. A., 111
Broken-wing displays, 88–92

Bronson, G. W., 2
Bubble nets of humpback whales, 50
Buchler, E. R., 78
Buchwald, J. S., 150
Bullock, T. H., 174
Bumblebees: foraging decisions, 62–63; dominance hierarchies, 187; adaptation for alfalfa flowers, 206
Bunge, M., 3–8, 42, 74
Burrows, animal, 95
Burrows, M., 174

Cable, C., 142
Caches, food, 69–71
Caddis fly larvae: cases, 98–101, 126–127; case repair, 100, 113; neural templates, 115
Callaway, E., 146
Canaries, responses to complex songs, 160
Carollia perspicillata, 187
Carter, G. R., 50
Caterpillars, behavior of, 33, 99
Cats, ERPs from, 150–152
Causation, ultimate, 22
Central nervous systems, 29, 144–153, 172–174
Cephalopods, brains of, 173
Cerella, J., 143
Cetaceans, 191–196
Chain communication by ants, 172
Chain reflexes, 35
Chalicodoma, 102, 103
Chapman, R. M. and J. A., 148–152
Chemical communication, 156, 172, 175
Cheney, D. L., 85, 167–168, 187, 208–209
Chickadees, 65, 71
Chimpanzees: termite fishing, 33, 125; self-recognition, 74–75; oddity learning, 140; courtship signals, 156; beliefs and desires, 191
Chisholm, A. H., 120
Chomsky, N., 2, 200
Churchland, P. M. and P. S., 3, 8, 31, 32
Clark's nutcrackers, feeding behavior of, 69–71
Clayton, D. A., 139
Clever Hans errors, 198
Cobb, J. B., Jr., 3, 29
Cognition: animal, 3, 19, 135; compared with consciousness, 135
Cognitive ethology, 12–15
Cognitive psychology, 11

Cohen, D. B., 203
Collias, N. and E., 108–109
Combinatorial productivity as criterion of language, 200
Communication: as "window" on animal thoughts, 38, 160–164; nonverbal, 38; by direction of gaze, 80; as evidence of thinking, 154–164, 208; incidental, 154–155; referential, 162–163; between parent birds, 169–170; in ants, 176; in honeybees, 176–185; of self-awareness, 206
Comparative psychologists, experiments of, 133
Compensatory behavior, 100
Complexity, as criterion of consciousness, 34
Concepts: 34; of pigeons, 142–143
Conditioned avoidance, 135
Conscious discrimination, ERP correlates of, 147–153
Conscious instincts, 41–47
Consciousness: animal, 1, 37, 47; relationship to language, 2; in infants, 2; brain mechanisms of, 3; definitions, 5; as awareness of own mind, 6; as illusion, 10; neglected by psychologists, 11; criteria of, 32 40, 47, 133; voluntary control of, 37; with a newly learned behavior, 41; adaptive economy of, 40–41; neurophysiology of, 44, 126; of genetically programmed behavior, 44–45; in absence of overt behavior, 45; as template matching, 116; in laboratory animals, 143; role in social cooperation, 159
Consensus, among honeybees, 183
Contingencies: of reinforcement, 21; of nutrition, 23
Cooperation: by ants, 106–107; by predators, 84–87
Cooperative hunting, 84–87
Corning, W. C., 6
Cowie, R. J., 71
Crabs: use of anemones by, 119; mating displays of, 158–159
Crane, J., 158
Crook, J. H., 108–109
Crows: mobbing, 87–88; dropping shells, 61–62, 121; olfaction, 53–54; searching images, 53–54
Croze, H., 53
Cuckoo bees, 103
Cullen, J. M., 65–66
Curio, E., 138

Daanje, A., 38
Darwin's finches, tool use by, 122
Darwin, C., 14, 49, 51, 63, 114
Davidson, D., 2
Davidson, J. M. and R. J., 3
Davies, N. B., 59, 65–66
Decisions by animals, 23, 40
Decoration, by bowerbirds, 128
Definitions of mental terms, 5–8
Delius, J. D., 141
Dennett, D. C., 2, 22, 29, 32, 121, 137
Descartes, R., 200
Desmedt, J. E., 146
Difflugia, 97
Digger wasps, tool use by, 120
Diptera, 119
Direction: communicated by honeybees, 179; by vervet monkeys, 167–168
Displacement: in animal communication, 164–185; in apes, 197
Displacement behavior, human, 109–110
Distance, communicated by honeybees, 179
Distraction behavior: gazelles, 81–82; birds, 36, 89–92
Diurnal rhythms, 137
Division of labor in social animals, 106, 169–170, 177, 188
Dogs: dreams, 203; wild, 84
Dolphins, thoughts of, 191–192
Domestication: effects of, 13, 14; by ants, 105
Donchin, E., 148
Dreaming: in animals, 202–204; EEG patterns, 203; eye movements, 203
Dreyfus, H. L., 11
Duerden, J. E., 119
Duncan-Johnson, C. C., 148
Dyal, J. A., 6

Earthworms, 48–49
Eccles, J. C., 3, 74
Ecologists, behavioral, 24–26, 52–58
Edwards, C. A., 49
Efficiency of conscious thinking, 40
Eisner, N., 174
Electrical correlates of thinking, 208
Electrical sensitivity of sharks, 53
Electric fish, communication, 154
Electroencephalograph recordings (EEGs), 144–145
Emotions, ERP correlates of, 152
Empathy: with animals, 16; between animals, 39

Environmental movement, 15
Environmental variability, in behavioral
  evolution, 52
Erber, J., 7
Ernst, U., 128
Escape tactics: of gazelles, 81; of vervet
  monkeys, 166
Evans, H. E., 101
Event related potentials (ERPs), 147–
  153
Evoked potentials, 145–153
Evolution, behavioral, 26, 51
Evolutionary biology, criticism of, 51
Experiments, as form of communication,
  163
Extrapolation, by honeybees, 206
Eye movements, 203

False incubation, 91
Falsification, importance of, 207–208
Fantasies of animals, 202–204
Farley, G. R., 150
Feeding behavior, 48–72
Feelings, neglected by psychologists, 11,
  21
Fellers, J. and G., 120
Female choice: 155–159; blackbirds, 55–
  58; weaverbirds, 108; bowerbirds, 127–
  128
Fenton, M. B., 76
Fiddler crabs, social displays of, 158–
  159
Field, T. M., 2
Finches, Galapagos: 63–65; tool use by,
  122
Fischer, W., 7
Fishbein, W., 203
Fisher, J., 67
Flexibility in invertebrate behavior, 173
Food caches, 69–71
Food gathering by birds, 58, 70–71, 169–
  170
Foote, S. L., 150
Foraging tactics: wagtails, 59–60; bumble-
  bees, 62–63
Foraminifera, skeletons of, 97
Fowler, H., 3
Fraenkel, G. S., 49
Free will, 7, 20
Frisch, K. von, 96–109, 128, 169–185,
  206, 209
Frustration, 136
Fullard, J. H., 76
Functionalism, 31–32
Fungus cultivation by ants, 104

Galambos, R., 3, 146–150
Gallistel, C. R., 204
Gallup, G. G., Jr., 74, 75, 205
Gardner, R. A. and B. T., 196–202
Gayou, D. C., 122
Gelman, R., 204
Generalization: experiments on, 138; gra-
  dients of, 138
Genetic influences on behavior, 21, 22,
  42–43, 51, 107–110
Glia, 30
Goldfish, visual generalization by, 141
Goldman, L. J., 78
Goodall, J. van Lawick, 33
Gorillas, signing of, 197, 199
Gould, C. G., 111, 112
Gould, J. L., 6, 111, 112, 180–181, 206–
  207
Grammar, in signing of apes, 199
Green, S., 157
Green jays, tool use by, 122
Grene, M., 173
Griffin, D. R., 3, 5, 76, 134, 161, 181,
  184, 205
Griffin, David R., 3, 29
Grizzly bears, 74
Gross, C. G., 144
Ground squirrels, alarm calls of,
  166
Gunn, D. L., 49

Habers, G., 141
Habituation, 91
Hagg, R., 193
Hain, J. H. W., 50, 191
Haith, M. M., 2
Hansell, M. H., 99–100
Harnad, S., 10, 12, 23
Hartman, E., 203
Hawks, attacked by smaller birds, 88
Haynes, B. D. and E., 74
Hediger, H., 3, 9, 75, 187, 205
Henson, O. W., Jr., 78
Herman, L. M., 50, 191
Hermann, H. R., 175
Hermit crabs, 119
Herons, use of bait by, 123
Herring gulls, shell dropping by, 60–61,
  121
Herrnstein, R. J., 142–143
Herron, J., 145
Higgs, M. D., 175
Hillyard, S. A., 3, 146, 147–150
Hinde, R. A., 67
Hölldobler, B., 170–173, 209

Honeybees: mating behavior, 158; symbolic communication, 176–185; foraging, 178; communication by queens, 181; swarming, 181–183; evaluation of cavities, 182; extrapolation, 206; innovative behavior, 206–207
Honig, W. K., 3, 135
Hopkins, C. D., 154
Hornbills, nests of, 169
Horridge, G. A., 174
Howse, P. E., 175
Hoyle, G., 77, 174
Huan, I.-N., 138
Huber, F., 174
Hulse, S. H., 3
Human reflexes, and instinctive behavior, 43
Human uniqueness, 1
Hume, D., 2
Humpback whales, 50, 139
Humphrey, N. K., 39, 159, 186, 187, 190
Hutton, R. S., 54
Hyenas: predatory tactics, 78–79; avoidance of familiar prey, 80; group predation, 84

Ideal free distribution, 56
Iersel, J. J. A. van, 102
If-then thinking by animals, 137
Images, searching, 53–54
Imitation, 138–139, 191–192
Inclusive behaviorism, 21, 38, 93
Individuality, animal, 18
Individual recognition, 165–166, 187–190
Information processing, unconscious, 12
Inherited templates, 114–117
Injunctions, alarm calls as, 168
Injury feigning by birds, 36, 89–92
Innovation by animals, 204–207
Inouye, D. W., 206
Inquiry versus assertion, 12
Insect behavior, 103–105, 111–113, 120, 176–185
Insects: food for blackbirds, 55–56; ultrasonic hearing, 76–77; neurophysiology, 77; nest repair, 115; possible thoughts, 170–185
Instincts: 42; conscious, 41–47
Intention movements, 37–38
Intentions: 5, 36–38, 120
Interdependence, of social animals, 38
Introspection, 10
Invertebrate animals: as robots, 173; as tool users, 118–120

Jackals, 81
Jackson, J. F. and J. H., 2
Janes, S. W., 120
Jennings, H. S., 31
John, E. R., 3, 10
Johnstone, J., 145
Jolly, A., 39, 159, 186
Jones, T., 122
Jouvet, M., 203
Jung, C. G., 46, 185
Jurasz, C. M. and V. P., 50

Kamil, A., 122
Kamiya, J., 203
Kandel, E., 7
Kawai, M., 67
Kenfoot, J. B., 74
Kenyon, K. W., 124
Kety, S. S., 5
Killdeer, antipredator behavior of, 36, 88
Kingbirds, antipredator behavior of, 87
Kin recognition, 187
Kinship, human-animal, 15
Kites, crayfish-baiting by, 123
Klemm, W. R., 144
Klosterhalfen, S. W., 7
Knee jerk reflex, 33
Knowledge, empathic, 7
Knudsen, E. I., 75
Koehler, O., 19, 204
Konishi, M., 75
Koslow, S. H., 146
Krebs, J. R., 24, 59, 71, 138, 176
Kroodsma, D., 139, 160
Kruuk, H., 78, 82, 84, 85
Kuhl, P. K., 2
Kutas, M., 145–148

Laboratory animals, limits of social behavior in, 13
Lacey, H., 21
Lack, D., 122
Lander, D. G., 143
Landmarks, marking hidden food, 70–71
Language: and consciousness, 2; contrasted with communication, 163
Leaf-cutter ants, 103–105
Learning: animal, 14, 20, 43, 140, 194, 206, 209; as criterion of mind, 6, 7, 42; compared to instinct, 45–46; oddity, 140; reversal, 140
Leger, D. W., 166
Leks, 158
Leopards, monkeys' alarm calls for, 166
Lepidostoma hirtum, 100

Lindauer, M., 173–188
Linden, E., 197, 198
Lions, hunting tactics of, 78–79, 86–87
Lloyd, J. E., 24, 173, 187
Loeb, J., 4, 33, 49
Lofty, J. R., 49
Logical positivism, 4
Lorenz, K., 3. 9, 37
Loveland, D. H., 142
Lovell, H. B., 123
Lowell, P., 207

Macaques, feeding tactics of, 67
MacKenzie, B. D., 10
Mackintosh, N. J., 20, 65, 135, 140
MacRoberts, M. H., 65–66
Maier, N. R. F., 6, 33, 119
Male display, 155–157
Markl, H., 173
Marler, P., 114, 139, 156, 157, 167, 208–209
Marshall, A. A., 128
Marsh hawks, auditory localization by, 75–76
Marsh tits, feeding behavior of, 71
Martial eagles, monkeys' alarm calls for, 166
Mason, W. A., 134, 138
Mason bees, nests of, 102–103
Masuhr, T., 7
Materialist philosophy, 5, 8, 44
McCray, J. W., 148–152
McMahan, E. A., 123
Mechanisms, animals as, 9
Mehler, J., 2
Melia tessellata, 119
Meltzolf, A. N., 2
Mental maps, 70
Mental states: 6–10; influence on behavior, 7; of animals, 136
Menzel, R., 7
Miles, H. L., 197, 199
Milk bottles opened by birds, 35, 67
Miller, E. H., 139
Miller, G. A., 200
Miller, L. A., 76–77
Millikan, G., 122
Mills, E. A., 74
Mind: animal, 3, 29; definitions, 5, 134
Mirror recognition, 205–206
Mishkin, M., 144
Mobbing, 88
Molecular regulation, 30
Monkeys: disappointment, 136; ERPs, 150–152; semantic and referential communication, 165–169; individual recognition, 187
Moods of animals, 38
Moore, M. K. and M. J., 2
Morrison, A. R., 203
Moths, evasive tactics of, 76
Motor learning, 36
Mountcastle, V. B., 149
Muskrats, 132

Nagel, T., 2
Names of animals, 205
Natural psychologists, 39, 186–190
Natural selection, difficulty of observing, 25
Nests: construction, 95; insects', 101–102; birds', 107–110
Netzel, H., 97
Neural mechanisms of consciousness, 30
Neural templates, 114–117, 131–132
Neurophysiology of consciousness, 44, 77, 126, 144–153, 174
Neuroptera, 119
Neville, H. J., 150
Nonverbal communication, 38, 161
Norman, D. A., 11
Norman, J. R., 97
Norton-Griffiths, M., 68
Novelty experiments with dolphins, 193
Number recognition by birds, 204

Object recognition by animals, 137
Observational learning, 66–69, 121–125, 137–139, 192
Octopus, brain of, 173
Oddball stimuli, ERP correlates of, 150–152
Oddity learning, 140
Oecophylla longinoda, 106–107
Oleson, J., 76
Olfactory components of honeybee communication, 180–181
Olsen, L. A., 74
Omitted stimulus, ERP correlates, 147
Operant conditioning, 141
Optimal foraging theory, 51, 59–62
Orangutans, signing by, 197–199
O'Reilly, J., 193
Orians, G. H., 55–58
Orioles, nests of, 107
Ornstein, R. J., 144
O'Shea, M., 174
Osmia bicolor, 102–103
Overtraining, effects of, 140
Owings, D. H., 166

Owls, auditory localization by, 75
Oystercatchers, feeding behavior of, 68–69

P300 waves, 147–153
Panpsychism, 29
Paramecium, 31
Parrots, communication by, 162–163
Parus, 65–66
Pattern recognition, 137, 142–144
Patterson, F. G., 197, 198
Pavlov, I. P., 4
Payne, R. S., 139
Pearson, K. G., 174
Pepperberg, I. M., 163
Peters, S., 139
Petitto, L. A., 199
Pets, human interest in, 16
Phenomenology, animal, 19
Pheromones, 175
Physicalist philosophy, 5
Picton, T. W., 147
Pigeons: surprise, 65; oddity learning, 140; color sequence learning, 141–142; operant conditioning, 141; person recognition, 142; object recognition, 142–143; concept learning, 142–143; concept of novelty, 193; number recognition, 205
Plovers: antipredator behavior, 88; predator distraction, 90
Polar bears, tool use by, 124
Polistes, 187
Pollen collection, 178
Poole, J., 143
Popov, A. U., 174
Popper, K. R., 2, 3, 40, 74
Porpoises: thoughts, 191–196; captured with tuna, 193; aiding behavior, 194–195; communication with, 195–196
Porter, F. L., 187
Potter wasps, nest repair by, 113
Predator-prey interactions, 73–94
Predators: distraction by birds, 36, 87–94; monitoring of, 79–80; alarm calls for, 166–167
Premack, D., 3, 191; and A. J., 184, 197
Prey monitoring, 82–84
Problem solving, 134–143
Protozoa, 30–31, 97
Pryor, K., 193–195
Psychiatry, 46
Psychology: cognitive, 11; experimental, 133–143
Purposive behaviorism, 42

Pye, J. D., 76, 164
Pyke, G. H., 62

REM sleep in animals, 203–205
Rachlin, H., 20
Rats: expectations of, 64; operant conditioning, 141; discrimination learning, 135
Ravens: dropping rocks, 120; number recognition, 204
Reductionism, 4, 24
Redwinged blackbirds, 54–58
Referential communication, 162–163
Reflexes, 33–34, 42
Reidinger, R. F., 138
Reinforcement, 21
Replication of experimental results, 14
Reversal learning, 140
Rice, W. R., 75
Ristau, C. A., 197, 198
Ritualization of aggression, 157, 171
Robbins, D., 197, 198
Roberts, G. J., 123
Robots, insects as, 110–114
Rocha-Miranda, C. E., 144
Rockstroh, B., 146
Roeder, K. D., 76
Roitblat, H. L., 3, 4, 22, 114, 134
Romanes, G. J., 14, 41, 114
Rose, S., 2
Rosenthal, R., 198
Ross, D., 119
Rowell, H. F., 174
Rue, L. L., 128
Rumbaugh, D. M., 3, 197–198

Saayman, G. S., 191
Sales, G., 76, 164
Sandpipers, 88
Sand wasps, 101
Savage-Rumbaugh, E. S., 3, 197–198
Savory, T. H., 9, 11
Schaller, G. B., 78–85
Schnierla, T. C., 6, 33, 119
Schwanzeltanz, 177–185
Schwartz, B., 21
Sea otters, tool use by, 124–125, 126
Searching images, 53–54, 70, 116, 192
Sebeok, T. A., 154, 198
Seeley, T., 182, 184
Seibt, U., 205
Self-awareness, 74, 75, 205–206
Self-concealment, as indication of self-awareness, 74
Semantic behaviorism, 134

Semantic communication: parrots, 163; monkeys, 166–169; ants, 169–176; honeybees, 176–185; apes, 196–202
Sensory components of evoked potentials, 146
Sentient nature, 15–17
Sexual selection, 155–157
Seyfarth, R. M., 85, 167–168, 187, 208–209
Shallice, T., 23
Sharks, electrical sensitivity of, 53
Shellfish-dropping, by birds, 60–62, 121
Shelters constructed by animals, 96–103
Sherry, D. F., 71
Shettleworth, S. J., 71
Shorter, J. M., 27
Signing by apes, 196–202
Silo pallipes, 99
Simplicity filters, 15
Skinner, B. F., 18, 19, 21, 141
Skutch, A. F., 36, 91
Sleep, EEG waves in, 145
Smith, W. J., 162
Snakes, monkeys' responses to, 166
Social communication in honeybees, 177–178
Social insects: complex behavior of, 170–185; as natural psychologists, 186–190; individual recognition by, 187–190
Solitary bees, 102–103
Songbirds, vocal imitation by, 138
Sordahl, T. A., 91
Species solipsism, 28, 161
Sperry, R., 3, 23
Spiders, 36, 111, 116
Squires, N. S., 150
Squirrels: recovering hidden food, 69; alarm calls of, 166
Stags, self-awareness in, 75
Starr, A., 150
Stenella attenuata and longirostris, 193–194
Steno bredanesis, 193
Stevens, R., 3
Stones as tools, 122, 125
Stotting, by gazelles, 81
Stoyva, J., 203
Straub, R. O., 141
Struhsaker, T. T., 166
Suarez, S. D., 75
Subjective experience, scientific study of, 11, 12
Surlykke, A., 77

Sutherland, N. S., 141
Swarming of honeybees, 181–183
Swencionis, C., 145
Symbolic communication, 165–185, 196–202
Symmetry, recognized by birds, 141
Synechdochaic fallacy, 11

Taboo, behavioristic, 20
Tabula rasa, concept of, 2
Tactics, animal, 24–26
Tactile communication, 176–179
Taxis, 33, 49
Tayler, C. K., 191
Templates, neural, 114–117, 131–132
Termite fishing: by chimpanzees, 33, 125; by assassin bugs, 123–124
Termites, nest repair by, 113
Terrace, H. S., 3, 4, 134, 141, 198–199
Territorial behavior, 158
Thatcher, R. W., 3, 10
Thinking: 3, 6, 8, 24–28, 35, 58; effect on behavior of, 21–24; adaptive economy of, 40–41; in predator distraction, 93–94; electrical indices of, 144–153; of apes, 201; wordless, 204
Thompson, R. K. R., 135
Thompson's gazelles, 79–82, 137
Thorpe, W. H., 6, 102, 108, 112–119, 120
Tinklepaugh, O. L., 136
Titmice, 35, 65
Tolman, E. C., 19, 42, 64, 136
Tool use by animals, 95, 118–127
Treat, A., 76
Treherne, J. A., 174
Trophallaxis in social insects, 177–178, 188–189
Tropisms, 33
Tucting, P., 146
Tuna fishing, and porpoises, 193–194

Uca, 158–159
Uganda kob, 158
Ultimate causes, 22
Umiker-Sebeok, J., 198
Unconscious behavior, 41, 46
Underwood, G., 3
Uttal, W. R., 3

Value judgments, 23
Vampire finches, 62–63
Van der Kloot, W. G., 99
Vander Wall, S. B., 69–71
Van Lawick-Goodall, 122

Versatility: as criterion of consciousness, 37; of animal behavior, 35, 108–110, 209

Vervet monkeys, communication of, 165–169; alarm calls of, 166–168; grunting sounds of, 168

Vieth, W., 138

Villers, P. de, 143

Vocal learning: songbirds, 138; humpback whales, 139

Voles, runways of, 96

Voluntary actions of animals, 7

Vultures, stone throwing by, 122

Waggle dances of honeybees, 177–185

Wagtails, feeding behavior of, 59–60

Walker, S., 3, 4, 20, 134–143, 170, 173

Wall, F. de, 156

Walther, F. R., 78

War, preparation for, and animal behavior, 112

Washoe (chimpanzee), 196–202

Wasps: nests, 101–102, 110; appropriateness of behavior, 110–112; thoughts, 117; tool use, 120

Wasserman, E. A., 3, 11, 20

Watson, J. B., 4, 18

Weaver ants: 106–107, 170–172; recruiting gestures, 171; chain communication, 172

Weaverbirds, nests of, 108–109

Weber, N. A., 104

Wenman Island, Galapagos, 63

West Eberhard, M. J., 101

Whales: feeding behavior, 49–50; brain, 50; songs, 139; thoughts, 191–196

Wheeler, W. M., 103

Whitehead, A. N., 29, 31

Whiteley, C. H., 10

Wiggins, G. B., 98

Wildebeest: as prey of lions, 82–84; failure to resist predators, 83; calves, 83

Wilder, M. B., 150

Wiley, R. H., 158

Williams, C. M., 99

Willows, A. O. D., 6

Wilson, E. O., 101, 103, 105, 170–175, 187

Wilson's plover, 90

Wilsson, L., 128, 131

Wittenberger, J. F., 24

Wolves: interactions with prey, 84; as beaver predators, 128–129

Woodfield, A., 5

Woodruff, G., 3, 184, 191

Word order, and grammar, 201

Worm-lions, tool use by, 119

Wright, W. H., 74

Yellowheaded blackbirds, 55–58

Yerkes, R., 197

Zach, R., 61

Zentall, T. R., 140

M